RANDOM NOISE

In the realm of safety management, this book embarks on a profound exploration of how the political economy was reshaped in the last two decades. Much like privatization, deregulation, and financialization altered the economic landscape, this narrative unveils how safety management has been affected by the intertwined dynamics of asset underinvestment, privatization, self-regulation, workplace flexibilization, and market-driven policies.

This book, the second installment of a thought-provoking trilogy on the consequences of neoliberalism, mirrors the political economy's promotion of the private sector's role in the economy. Just as neoliberalism amplified and accelerated the mechanisms of human-made disasters in complex systems, this narrative lays bare the heightened potential for safety misfortunes when governed by market-driven principles.

As the story unfolds, the book delves into the concept of 'synoptic legibility' in safety management, akin to how the political economy distilled its essence into privatization and deregulation. The authors scrutinize the consequences of translating safety measures into rigid targets, unveiling how this shift can distort the integrity of safety metrics and inadvertently harm individuals. Drawing parallels to historical blunders such as England's window tax, the book contemplates the precarious nature of equating simplified metrics with safety achievements. Much like the political economy's 'acceptable risk' renegotiations, it examines how the pursuit of safety through metrics and surveillance can lead to 'manufactured insecurity,' eroding trust, autonomy, and professionalism.

In *Random Noise*, Poole and Dekker extend this reach once again, writing for all managers, board members, organization leaders, consultants, practitioners, researchers, lecturers, students, and investigators curious to understand the genuine nature of organizational and safety performance.

Georgina Poole is a Health and Safety Professional in Brisbane, Australia. She has spent more than 15 years embedded in operational safety roles across a variety of industries, including mining, oil and gas, construction, and aviation. Georgina is well known for her podcast 'Leading Safely' and her passion for human and organizational performance. She recently completed her Graduate Certificate of Safety Leadership at Griffith University, where co-author Sidney Dekker is the Program Director.

Sidney Dekker is a Professor at Griffith University in Brisbane, Australia, where he founded the Safety Science Innovation Lab. He is also Honorary Professor of Psychology at the University of Queensland. Previously, Dekker was Professor at Lund University in Sweden, where he founded the Leonardo da Vinci Laboratory for Complexity and Systems Thinking and flew as First Officer on Boeing 737s for Sterling and later Cimber Airlines out of Copenhagen. Dekker is a high-profile scholar and is known for his work in the fields of human factors and safety. He is the author of many best-selling titles, such as *Just Culture*, *Safety Differently*, *The Field Guide to Understanding Human Error*, *Drift into Failure*, *The Safety Anarchist*, and *Foundations of Safety Science*.

THE BUSINESS, MANAGEMENT AND SAFETY EFFECTS OF NEOLIBERALISM

This mini-series sets out to explore the far-reaching effects of neo-liberalism. Neoliberalism supports and promotes the role of the private sector in the economy, and argues that markets and free trade are the only way to get the best price for the best quality. In this series, Sidney Dekker examines how neoliberalism actually affects business, management, and safety in practice.

Compliance Capitalism
How Free Markets Have Led to Unfree, Overregulated Workers
Sidney Dekker

Random Noise
Measuring Your Company's Safety Performance
Georgina Poole and Sidney Dekker

For more information about this series, please visit: www.routledge.com/The-Business-Management-and-Safety-Effects-of-Neoliberalism/book-series/BMSEN

RANDOM NOISE

Measuring Your Company's Safety Performance

Georgina Poole and Sidney Dekker

Routledge
Taylor & Francis Group

LONDON AND NEW YORK

Designed cover image: tiero/Getty Images®/iStock collection

First published 2024
by Routledge
4 Park Square, Milton Park, Abingdon, Oxon OX14 4RN

and by Routledge
605 Third Avenue, New York, NY 10158

Routledge is an imprint of the Taylor & Francis Group, an informa business

British Library Cataloguing-in-Publication Data
A catalogue record for this book is available from the British Library

Library of Congress Cataloging-in-Publication Data
Title: Random noise : measuring your company's safety performance /
  Georgina Poole and Sidney Dekker.
Description: Abingdon, Oxon ; New York, NY : Routledge, 2024. |
  Series: The business, management and safety effects of neoliberalism |
  Includes bibliographical references and index.
Identifiers: LCCN 2023055591 (print) | LCCN 2023055592 (ebook) |
  ISBN 9781032012438 (hardback) | ISBN 9781032012421 (paperback) |
  ISBN 9781003177845 (ebook)
Subjects: LCSH: Industrial welfare—Moral and ethical aspects. |
  Industrial safety—Management. | Industrial safety—Economic aspects.
Classification: LCC HD7261 .P6185 2024 (print) | LCC HD7261 (ebook) |
  DDC 658.3/8—dc23/eng/20240104
LC record available at https://lccn.loc.gov/2023055591
LC ebook record available at https://lccn.loc.gov/2023055592

ISBN: 978-1-032-01243-8 (hbk)
ISBN: 978-1-032-01242-1 (pbk)
ISBN: 978-1-003-17784-5 (ebk)

DOI: 10.4324/9781003177845

Typeset in Joanna MT
by Apex CoVantage, LLC

CONTENTS

ACKNOWLEDGEMENTS

We want to acknowledge Josh Bryant for his ongoing support and advocacy. His work operationalising human and organisational performance principles and safety differently is truly inspirational. We would also like to thank Laurin Mooney, Ken Madson, James Newman, Brian Hughes, Jon Schmidt, Andy Barker, and Carter Pittard for being there to read, listen, and share and Darrell Horn for his indefatigable editing and improvement of the manuscript.

PREFACE

Lost-time injuries (or LTIs) were never meant to do what many organisations are making them do today. Organisations and entire industries typically equate the number of LTIs or their frequency across hours worked with safety. But LTIs never were a safety measure. They are a measure of labour efficiency. In pre-industrial labour relations, people generally purchased a worker's actual labour. Or they directly bought the product produced by that labour. But as we went through the Industrial Revolution in the late 19th and early 20th centuries, the means of production were concentrated with (capitalist) owners. The labour to operate those means of production was sourced through large contingents of workers. Productivity and prices now had to be measured differently.

Capital purchased a worker's labour *time*, or *potential* labour, instead of products or actual work. In such an arrangement, pursuing strategies that regulated the labourer's productive processes became natural. The point was to derive as much work, and thus value, as possible from a given amount of purchased labour time (Newlan, 1990). Meeting the needs and problems of the 20th century was a new type of management: 'scientific management.' Its best-known proponent was, of course, Frederick Taylor. In testimony before a Special House Committee of the US Congress in 1912, Taylor expressed that true Scientific Management requires a mental revolution on the parts both of management and of workers. . . . [T]he interest of both and of society in the long run call for ever greater output of want-satisfying commodities. Output requires expenditure of human and material energies; therefore both workers and management should join in the search for discovery of the laws of least waste.

(Taylor et al., 1926, p. xiii)

Pursuing 'least waste 'in an industrialised economy full of potential labour made good sense. In a way, it was getting the most significant 'bang for the buck.' Injuries that led to lost time meant that potential labour was wasted. Like everything in scientific management, this waste had to be quantified and managed. As an owner or operator of the means of production, you would want to know how much you were 'wasting' or whether you were getting value for money from the potential labour you had purchased. So a measure of labour productivity came in handy. Lost-time injuries (LTIs) and medical treatment injuries (MTIs) directly reflected lost productive capacity and reduced surplus value in such an industrial arrangement. Lost-time injuries refer to workplace injuries or incidents that result in an employee's inability to work for a certain period, typically a shift or more, due to the severity of the injury or the time required for recovery.

Tracking and analysing workplace injuries gradually became a proxy measure of safety when factories and industrial operations began to see an increase in accidents and injuries. Workplace safety was often overlooked before the mass accumulation of labour under single factory roofs. Accidents were considered an inevitable part of hazardous working conditions. But as industrialisation grew, so did the visibility of accidents (Burnham, 2009) and the concomitant concerns about worker productivity. Lost-time injuries could morph into a proxy safety metric because they represented incidents (and physical consequences) severe enough to warrant time off work.

As organisations became more aware of the financial costs associated with workplace injuries, they began to see the value in tracking and reducing lost-time injuries. High rates of lost-time injuries resulted in direct costs related to medical expenses and compensation and indirect costs due to productivity losses, service disruptions, increased insurance premiums, and potential legal implications. The human cost started to bother some as well. In the mid-20th century, the focus on occupational safety continued to intensify, leading to the establishment of regulatory bodies and standards

to protect workers. These standards not only required organisations to provide safe working environments but also encouraged the implementation of
measures to reduce workplace injuries.

Today, LTI and MTI, as essentially cost and productivity figures, are used
as stand-ins for a lot of other things: workplace safety, injury frequency,
injury severity, workplace culture, national safety culture, workplace health
and safety costs, and even individual worker performance or suitability for
the job (O'Neill et al., 2013):

> Rather than offering a measure of that subset of injuries indicative of lost
> workplace productivity, corporate reporters are increasingly presenting LTI
> numbers as measures of (total) injury performance and even of occupational
> health and safety itself. Critics suggest injury data routinely forms the corner-
> stone of occupational health and safety performance reports with an almost
> exclusive status quo reliance on recordable and lost time injury rates as safety
> performance measures.
>
> (p. 185)

Over time, concerns have arisen about the limitations of using lost-time
injuries as the sole or primary measure of safety performance – or as a
safety measure at all. The metric can encourage under-reporting of injuries
to maintain favourable statistics, leading to a skewed perception of safety.
Additionally, focusing solely on lost-time injuries might not adequately capture the safety culture and proactive measures organisations take to prevent
accidents.

It turns out that the problems and myths associated with equating LTIs
with workplace safety are more profound than many might think (Besnard
& Hollnagel, 2014). This book will walk you through them and give you
something better to work with instead.

Chapter 1 sheds some light on the lure of homogenised quantity. Why
is it so attractive to think we can measure something complex with a
single metric? Where does that illusion come from? We trace it back to the
aspirations of the Enlightenment and the introduction of 'insecurity' to
the English lexicon, simultaneous with the promises of modern industrial
means and scientific measurement to control and contain that insecurity.

In Chapter 2, we plumb various critiques of injury measurement to show
that the LTI emperor has no clothes. Statistically, the figure is nothing more
than random noise. To make any managerial claims about a drop in LTIs

year over year or to worry that one LTI can somehow be traced back to managerial decisions is as hubristic as it is utterly ignorant. The chapter will have to do some statistics, but it makes the overall argument quite accessible.

Chapter 3 brings that argument fully in from the storm. It explains that the likelihood that a reduction in the number of injuries is caused by the intervention rather than by chance is statistically significant. If you want to be confident that a reduction in injuries is due to what you did rather than mere random variations (which can go as wildly up as they can go down!), the numbers need to meet various stringent requirements. Spoiler alert: Your organisation's data will fail to meet those requirements.

Chapter 4 outlines how the absence of injuries doesn't predict what you might think it does. It has been popular, ever since Heinrich in the 1920s, to believe that low injury numbers are somehow an assurance that you won't suffer any worse either. Much research since then shows that this is untrue. In the most likely case, the relationship between injuries and accidents is random (or non-existent). In some cases, researchers even suspect that the absence of (reported) injuries is an omen that suggests an accident may well be around the corner.

Chapter 5 takes you to the typical victims of severe or fatal incidents and accidents in the workplace: your contractors. It traces the research explaining how and why contractors appear at greater risk of these life-changing events than your employees. Again, some statistics are required, but you probably don't need them to know where you might start looking for safety gains.

Chapter 6 provides insights into the 'ultra-safe' aviation industry focusing on research based on the corporatization of air navigation systems in Norway. The research clearly demonstrates linkage between variations in perceptions of safety during periods of organizational change.

Chapter 7 provides a deep dive into the complexity of issues relating to the accuracy and reporting of accidents and injuries within organizations. This chapter looks at how elements such as supervisor enforcement and safety climate play a part in how comfortable an employee is sharing their story when an accident or injury has occurred.

Chapter 8 challenges the perennial notion that all injuries and incidents are preventable. It runs – briefly – through arguments from complex science to show that the world is not so easily fooled. Then, it turns to the debates surrounding 'zero harm' as an expression of the total-preventability belief. In contrast, humanity has long committed to a desire to ban suffering (as many religious beliefs also propose), but practical achievability remains elusive.

Chapter 9 then rightly asks what we should be measuring instead. It takes us on a brief tour of where safety science has been turning since the full-scale disappointment with LTIs (and other measures of the absence of adverse events) became apparent. Instead of seeing safety as the absence of bad events, safety is the presence of capacities that make things go well. This is the basis for safety differently, safety II, and resilience engineering. The chapter talks about the sorts of capacities that these schools of thought suggest we measure instead.

Chapter 10 wraps up the book by turning directly to the people in your organisation who need to make these things happen. What do your safety people have to do now? If LTI is no longer their role, what should they spend their time on? Research tells us your best bet is supporting frontline capacity building and adaptations. The research outlines specific activities for safety professionals to create foresight about the changing shapes of risk and how they can help identify and enhance the capacities that make things go well in your organisation.

This is the second book in a three-part series. Part 1 described how it is possible that free markets, with less government 'interference' in how business is done, actually lead to more compliance burdens and unfree workers. It showed, among many other factors, how financialisation helps explain organisations' preoccupation with short-term performance because that is what markets hold them accountable for. The quarterly reporting demands get discharged by constantly chasing a range of indicators, metrics, and benchmarks – including injuries or injury rates. Many of these measures have stopped being measures that help managers manage. Instead, the measures become objects to be managed to drive specific market outcomes.

This second book in the series picks up on that point. It shows how the injury numbers used – particularly by organisations for quarterly reporting and boards for liability management and due diligence protectionism – are nothing more than random noise. It will explain how reliance on these numbers can amplify and accelerate the drift mechanisms into failure and human-made disaster. It argues how the increased potential for system collapse lies beyond the predictive reach of low numbers of adverse events – and that we should be identifying and enhancing the capacities that gener-ally make things go well instead. We can even develop ways to measure these capacities and provide metrics to those who believe their organisational roles, bureaucratic mandates, or governance accountabilities demand them.

The third book in the series then focuses on the immense amount of compliance clutter stemming from the contorted, unchecked workings of the supposedly free market itself. It will do so by invoking the idea of freedom in a frame and reintroducing trust and professionalism. These are still more recognisable in what is known as Rhineland capitalism (as opposed to Anglo capitalism), which acknowledges that complexity can never be harnessed or governed through rules and compliance. Instead, it requires horizontal coordination, adaptation, expertise, and trust. It offers a vision of humanity and organisational life, which Anglo-inspired capitalism will have great difficulty offering if left to its own devices.

1

THE LURE OF A HOMOGENISED QUANTITY

In the early 1700s, King William III of England imposed a new, banded, progressive property tax. Like all progressive taxes, more meant more. That is, the bigger or more valuable your asset was, the higher your marginal rate on its value relative to others' assets would be. Progressive taxation seems fair in principle, but it requires that the bands be differentiated and that you know when you go from one to the next. Lines need to be drawn between the bands. But how precisely do you determine the value of an asset? Where exactly does the line go and based on what? This can be troublesome even if that asset is real estate. The solution was to hone in on a simple metric: a homogenised quantity. What's the easiest thing we can count, even from the street? With no need to even go inside? You may have guessed. In 1747, the English government had found the solution:

- For a house with 10 to 14 windows, the tax was six pennies per window (about half a dollar in today's money).
- For a house with 15 to 19 windows, it was nine pennies.

DOI: 10.4324/9781003177845-1

- For a house with twenty or more windows, it was one shilling (or twelve pennies) per window.

In search of a homogenised measure, the English started counting what they could count. One could argue that the number of windows was a fair proxy for the size of the property. Therefore, it would be a fair proxy-proxy to the property's value. (Though, as we know, the location of a property can mean even more.) But the English had other ideas. Their answer to this taxation was not to pay the tax but to do away with the windows that led to them having to pay it in the first place. The English started restricting both the number and size of the windows in their houses. It didn't help their health or hygiene, of course. Their neighbours across the Channel had done the same not long before:

> Officials of the French absolutist kings sought to tax their subjects' houses according to size. They seized on the brilliant device of counting the windows and doors of a dwelling. At the beginning of the exercise, the number of windows and doors was a nearly perfect proxy for the size of a house. Over the next two centuries, however, the 'window and doors tax,' as it was called, impelled people to reconstruct and rebuild houses so as to minimize the number of apertures and thereby reduce the tax. One imagines generations of French choking in their poorly ventilated 'tax shelters.' What started out as a valid measure became an invalid one.
>
> (Scott, 2012, p. 115)

What gets measured gets manipulated, particularly if the measure is connected to a target, such as a maximum number of windows in a house so that you don't click over into the next band. The setting of a target number makes manipulating the measure even more acute. Once a measure becomes a target, it stops being a measure. It becomes a target, and people adjust their behaviour to meet it. The measure of taxable property in this example stopped being a measure. It started driving a target, and it even became a target. People adjusted their behaviour to meet it.

A measure with a target stops being a measure

Two effects are noteworthy. First, whatever the measure was, it stopped being meaningful. Precisely because it had been a proxy for the size (and thereby a proxy for the value of the property), a deliberately reduced number of windows and doors no longer represented the value of the taxable asset. Continued application of the measure automatically kept reducing its validity. Second, the behaviour driven by the measure/target undermined the validity of the measure itself and triggered secondary consequences contrary to any emerging nation's goals: an onslaught on the health of its people. Both these effects led to the repeal of such measures in the 19th century, not only in France, but in Scotland and England. Campaigners argued it was a light and air tax and, thereby, a population health tax.

The Enlightenment and early modernism supply ample examples. We got progressively better at measuring things and understood the power of what we were learning and what we could manipulate. During the Enlightenment and modernism, Western society dramatically transformed from what it had previously been. Liberation from the dictates of kings or Church, followed by industrialisation and urbanisation, gave us confidence in the power of rationality, planning, measurement, and science. Armed with these, we could better control our health, our livelihoods, our economies, our productivity, our systems, and our people.

Synoptic legibility

A strong desire for those in governance or managerial roles it to get a sense of what's happening in their system: a one-glance view of the state of some vital parameter. The number of windows in a house would be one example. The more industrialised, bureaucratic, or institutional the governance is (i.e. the more faceless and distant it is from people's actual, nuanced, messy lived experiences), the more this pursuit of a homogenised quantity might seem to make sense. You count what you can count, even if it doesn't count and even if it creates all kinds of other undesirable effects:

> According to Marcuse (1991), what is at stake is the compatibility of technical progress with the institutions in which industrialisation developed and vice versa. A spectre haunts these institutions and society, that perhaps they are both products of the twentieth century with no place in the next. Can they change and confront the challenge, or will they hang onto the past and face extinction? Too often, modern institutions/societies respond to this uncertainty with bravado and the ritual (thought-less) application of pseudo-scientific procedures and methods. Audits, analyses, surveys, systems

and management reviews abound. . . . The gathering of information provides a ritualistic assurance that the appropriate attitude about decision-making exists.

(Angell & Straub, 1999, p. 196)

What organisations and institutions want, ideally, is 'synoptic legibility.' Synoptic legibility makes something too complex and challenging to wrap your arms around readable. It provides the kind of ritualistic assurance that Angell and Straub talk about. It turns the complexity of an organisation into a single metric that is understandable from a single perspective or viewpoint – without even having to be in or at the place of measurement. You can know about the value of a property by staying on the street. You can know about safety by getting an LTI number. You don't have to go on-site. You don't have to dirty yourself at the frontline to understand how work is actually done. To make monitoring safety from a distance quick and standardised, an LTI is perfect. It creates a metric, a view, that gives anybody a standardised summary of a particular characteristic of an organisation or industry: one that you can easily compare with other operations, companies, or industries.

The idea of synoptic legibility comes from the period of Enlightenment as well. However, it brings back the pre-Enlightenment sovereign desires for control that the Enlightenment was supposed to leave in the past. Between 1853 and 1869, Baron Georges-Eugene Haussmann, a French civil engineer, was tasked by Louis Napoleon to undertake a vast re-engineering of Paris. Haussmann created the street grid that we know as Paris today. The lack of legibility was one of the key inspirations for the grandiose project. The transformation's goals were simplification, legibility, straight lines, rationalisation, central management, and a synoptic grasp of the whole. Though these were visible at the surface, wider streets and boulevards to celebrate the Second Empire's glory weren't the most important aims. The main goals were control and security. Louis Napoleon had seen revolutions in 1830 and 1848, and resistance to his rule was still concentrated in dense, messy Parisian working-class quarters. These evaded effective surveillance. Haussmann's Paris was designed to map and gain control over this *ceinture sauvage* (literally, 'wild ring'). The redesign of these neighbourhoods made them more legible, and the street plans now connecting them allowed Louis Napoleon to pour soldiers into the revolutionary areas swiftly.

Self-harm and suicide by prison inmates is a more recent example. Western countries have reported this to be a growing problem. In the

15 years from 1972 to 1987, the United Kingdom saw a sharp increase in prison suicides. In Finland, almost half of all prison deaths result from suicide. Hanging is a method often used. The problem for prison administrators is not just a humane one. It is a liability problem and a reputational one. Suppose it can be demonstrated – in hindsight, of course – that prison officials were indifferent to the fate of a prisoner with suicidal tendencies. (Those with a history of mental health problems are more likely to end up in that category.) In that case, they can be held liable for withholding medical care. Prison deaths are also a public relations problem for politicians and tough-on-crime governments. As a result, one Western country fines its prisons $100,000 for each suicide within their walls.

This is where the suicide watch comes in. Suicide watch is meant to ensure the inmate's safety and protect the prison and its officials against liability (or, indeed, against hefty fines from the government). Inmates under suicide watch are placed in an environment where it is difficult for them to hurt themselves. Any objects that could be used in self-harm are removed (furniture, fittings, hooks, door-closing brackets, bed sheets, rigid walls, belts, laces, neckties, shoes, socks, suspenders, tampons). In many cases, nothing but a padded cell is left, with a nude inmate in it. The light is left on 24 hours a day. In even more extreme cases, inmates can be physically (or, to use the euphemism, 'therapeutically') restrained. Chemical restraint (with sedative drugs) is a last option.

All humanity, social interaction, cognitive stimulation, and human dignity are removed. Human harm is prevented by imposing profound human harm. If what gets measured is the number of suicides, and a price is put on each one, then that number needs to be as low as possible. A measure (number of suicides) has become a target. (We want zero suicides.) Any desire of the incarcerated to commit suicide may well be exacerbated by official efforts to avert it. Other institutions driven by specific outcome numbers display the same global inefficiencies and dysfunctions. In discussing how this plays out in a hospital setting, Cook and Rasmussen (2005) describe a condition they dubbed 'going solid.' This is the process of a system shuddering to a halt and activities within it congealing because of the focus on locally rationalising and optimising the few things that can be counted:

> 'Going solid' generates new performance demands. Clinical decision-making becomes pressurised as practitioners struggle to identify the patients most likely to do well with lower levels of care to create openings for new patients.

The average workload within units rises as the least sick are moved out to accommodate those who are sicker. The situation is ripe with perverse incentives for practitioners to 'game' the larger system in order to manage workload. By overstating the acuity of patients in a unit, nursing workload measures may be manipulated in order to garner nursing resources. New patient admissions may be delayed by misrepresenting the bed availability in that unit – for example, by claiming the bed location is being 'cleaned.' Some of the gaming can be quite subtle. A practitioner may 'block' an ICU bed by keeping it occupied by a patient ready for discharge to the ward, timing the discharge of that patient in order to recapture the bed for another patient – for example, for an elective surgical procedure that will require an ICU bed for postoperative care. Paradoxically, although the incentives to hide resources in this way arise because the solid system has little available 'slack,' these locally effective strategies perpetuate the solid condition.

(p. 132)

Managing the measure

Measuring a countable quantity (like available beds) or a negative (such as the number of injuries) incentivises efforts to produce or stage-manage a particular measurement. Pressures to carefully 'case-manage' injuries (see the example in the next chapter) have been widely noted. Contractors or providers who do not play along may see their contracts ended (Tozer & Hargreaves, 2016). The history of loss-time injuries (LTIs) mimics that of taxable windows in houses, suicides in prisons, and beds in hospitals. It never started as a safety measure. Then, it was turned into one. Then it became a target. And as the measure started to get manipulated to meet the target, it lost any connection to safety. It started sacrificing the health, well-being, and safety of entire organisations and their workforces. This happened not only through the unending clutter of reporting, recording, and procedural interventions but also by actively creating more significant risks of accidents in an organisation that had become dumber and pervaded by risk secrecy (e.g. Hopkins, 2010).

Though it can benefit a few others – for a little bit, for example, a Louisiana man had to spend time in prison for lying about worker injuries at a local power utility, which allowed his company to collect $ 2.5 million in safety bonuses. A federal court news release says the 55-year-old was sentenced to serve 6.5 years in prison, followed by 2 years of supervised release. He was the safety manager for a construction contractor, convicted of not reporting injuries at two different plants in Tennessee and Alabama between 2004 and 2006. At his federal trial, jurors heard evidence of more than 80 injuries that were not promptly recorded, including broken bones; torn ligaments; hernias; lacerations; and injuries to shoulders, backs, and knees. You, of course, wonder about the fairness of jailing a middle manager rather than going higher up. But, as always, the evidence becomes more diluted the further up the hierarchy you go looking. In any case, the construction contracting company did pay back double the bonuses (Derango, 2013). There are other examples as well, including this one from international shipping:

> For instance, establishing indicators for monitoring rest hours and the unwillingness to supply additional manpower when rest hours are not being met, setting up unrealistic goals such as zero accidents and expecting openness from the crew in reporting accidents, assigning ambitious timeline to accident reports and turning to an under-resourced safety department to close reports in time, and setting up uncompromising deadlines for the maintenance of safety-critical equipment with minimum spare parts and time allocated for maintenance. What follows is a deliberate manipulation of metrics, such as falsification of rest hours, a culture of fear and underreporting of incidents, questionable quality of accident investigations, and a deferral of maintenance based on risk assessments and waivers to meet individual and departmental key performance indicators. Managing the measure takes precedence over measuring to manage.
>
> (Anand, 2016, p. 21)

The window tax and suicide watch pattern are repeated in all this. First, the measurement stopped being a measurement because it became a target. Because it became a target, it no longer measured what it was meant to because people started to manipulate ('case manage') their numbers to meet the target. Second, it triggered secondary consequences that harmed employees. This harm was not just inflicted directly: for example, by bullying employees into wearing the yellow vests or denying the reality of their suffering by renaming and not counting their incidents. Harm, or potential harm, was also imposed indirectly and more widely by not investigating the sentinel events that could give rise to more significant disasters. The resulting cultures of risk secrecy and obfuscating, renaming, and euphemising harm directly counter an organisation's and industry's safety goals. Some governments have been waking up to the problem. Responding to concerns that internal safety management systems were, in some cases, likely to suppress injury data, worker dissent, and other bad news, the US Government Accountability Office sent a report to Congress demanding better occupational health and safety guidance on safety incentive programs and the kinds of counterproductive measurements they promote (GAO, 2012).

Manufactured insecurity

Safety as a single digit has become a way of thinking about what goes on in an organisation in and of itself, even if this is disconnected from the supposedly 'actual' work going on there – the core activities, systems, and processes; the work; and the people who co-produce safe outcomes through their operations. Safety as a single digit has become an increasingly self-referential datum, consistent primarily with itself and independent of the complex, dynamic world to which it purports to speak. That it would come to this is perhaps not so strange. As Townsend explains:

> 'Safety' is talked about as if it were a physical or social entity that, by impli-cation, can be studied in the same way as a physical or social science. At its best, the study of safety is about understanding that, as time progresses, rates of failure decrease and about trying to interpret the processes involved. At its worst, the study of safety is used to chauvinistically justify a particular way of doing things to the exclusion of others; improving failure rates are used to imply success by association without the admission that there might be other explanations. In the absence of tangible objective measures, safety is 'measured' using subjective surrogates.
>
> (2013, p. xiv)

LTIs make work productivity (and, putatively and by proxy, safety) accessible and legible. It is a hugely subjective (and meaningless) surrogate, but it suggests, in Townsend's words, that there is a tangible objective measure. Through LTIs, the safety of work becomes an ephemeral representation, a single digit. And it is a single digit directly connected to the security of people's positions up and down the organisational hierarchy. Townsend again:

> Industry and the people who run it are under ever-increasing public expectations and legislative pressure to improve health and safety. They are expected to continuously improve something in which improvement cannot be measured.
>
> (2013, p. xiv)

Those on the frontline typically recognise that improvements cannot be measured easily, if at all. They have an acute sense of the uselessness (or even counterintuitive or fraudulent nature) of the kinds of health and safety measures used to try to make their daily reality synoptically legible. The reported and recorded metrics say nothing about the obstacles and frustrations they encounter on every shift or the nuances and messy details of their department's safety-critical processes. Even leaders occasionally recognise that they need to understand how work is actually done under actual challenging conditions – to 'get out on the decks' to understand what is actually going on:

> [A]pproximately eight months before the Macondo blowout, Transocean President Steven Newman forwarded his observations about Transocean's use of leading indicators to several senior Transocean managers: 'I am not convinced at all that we have the right leading indicators. The leading indicators we report today are all just different incident metrics – they have nothing to do with actually preventing accidents. . . . [T]he only way [we] could really meaningfully answer the questions would be to get out on the decks.'
>
> (CSB, 2016, p. 148)

The demonstration of improvement in the number of lower-consequence events hardly yields the insight necessary to prevent significant events, as we will discuss in more detail later in the book. In industries that show near-zero safety performance (i.e. they have only a small number of incidents or injuries), the predictive value of these incidents for larger-consequence accidents (including fatalities) declines dramatically as well – if it doesn't invert (Mendelhoff & Burns, 2013; Saloniemi & Oksanen, 1998). The more incidents you have (or are honest about), the fewer significant accidents or fatalities you might suffer. Research on zero-harm policies shows the same inverse relationship (Sherratt & Dainty, 2017), as do many empirical cases in the public record. The Texas City refinery and the Macondo Well had been celebrating low numbers of injuries and incidents right before a fatal process catastrophe. As observed by Amalberti (2001):

> All this additional information does not necessarily improve the prediction of future disasters. The logic behind the accumulation of data first relied on the strong predictability quasi-accidents had on accidents; extending the scope of safety analysis to quasi-accidents seemed natural. The same logic then applied

by linear extrapolation to incidents, then to quasi-incidents, and eventually in turn to precursors of quasi-incidents. The result is a bloated and costly reporting system with not necessarily better predictability, but where everything can be found; this system is chronically diverted from its true calling (safety) to serve literary or technical causes.

(p. 113)

Despite the impossibility of measuring an improvement, organisations try to do so anyway, creating recording and reporting systems which, in Amalberti's words, serve literary or technical causes (or, in other words, fulfil bureaucratic accountability requirements and drive or justify managerial interventionism). Organisations can even use the records inversely – claim that these represent or measure a deterioration – which then enables them to hold someone to account for them. A supervisor can get in trouble for allowing even a single LTI to show up in their work area. A manager can be excoriated for having a worse LTI record than colleagues at other sites or business arms. Senior leaders can be made to pay for the insecurity of worsening LTIs by losing bonuses or worse. Company boards may feel the insecurity of LTIs in the possibility of their being challenged on their due diligence. How publicly listed organisations are held accountable actively contribute to this state of affairs:

The quarterly report syndrome is proper to this approach. There must be constant and immediate signs of success, and these must be structured to encourage the stock market and throw constant sops to the board of directors. Long-term planning, fundamental and long-term investment . . . to improve . . . are the last things they want.

(Saul, 1993, p. 365)

However, gaining security by pursuing low (or zero) LTI will inevitably be counterproductive. It may take a while, but the deleterious effects will appear. The main mechanisms by which managers are told to gain security (get the LTI down, keep it down!) involve investing in a system with the main ingredients necessary for LTIs to go up. These include risk secrecy, eroded safety cultures, under-reporting, a lack of integrity and honesty, undermining trust, and a sapping of engagement (Dekker, 2022). The response is typically to engage in more surveillance, reporting, and recording with cameras; intelligent vehicle-monitoring systems; drones; wristbands;

and tracking devices. This 'machinery' for the surveillance and monitoring of human behaviour is primarily accepted and hard to resist up and down organisational hierarchies (Harrison & Dowswell, 2002; O'Loughlin, 1990). Cultural conditioning, drifting societal norms, and the expectations of the °stock market are all legitimate excessive attempts at control. As Amalberti explains:

> Problems surrounding safety in our society have never been so widely discussed. . . . [We] have responded by creating agencies, new authorities and offices dedicated to safety; these bodies have passed numerous laws and decrees on safety; research funding has flowed rapidly into this area (research funding in this area in Western countries has increased by almost 300% since 1992). The same inflationary tendency is seen in university and continuing training courses, while small and medium-sized enterprises and specialist consultancies in this niche of safety intervention and consultation are becoming more numerous (+72% in the United States between 1985 and 2000) to meet the exponential demand for audits and risk assessment reports (+430% in 10 years). It is this emergence of the concept of 'safety' which has finally given scientific credence to a subject which was long treated simply as a variable associated with technical development. Not every aspect of this public awareness is beneficial, however, the market that has been created is not only lucrative but it also has a polemical streak and still suffers from a number of vulnerabilities.
>
> (2013, p. 2)

The organisational life this has given rise to is all too familiar for many. Surveillance and the attempts to reduce uncertainty and insecurity are driven by and require bureaucratic accountability, rules and procedures, systems and processes, administrative back offices, and IT infrastructures to furnish and maintain them. These now embody what Foucault called *governmentality*: a complex form of power that links individual conduct and administrative practices. People may recognise surface features of governmentality today as 'safety clutter' (Rae et al., 2018). It amounts to a relentless addition of rules, procedures, systems, and processes surrounding the capture, reporting, tabulation, tracking, ranking, storage, and analysis of all the data that comes with governmentality. But they are all expressions of distrust, of an acknowledgement that honesty has already left the building. Control and coercion are what is left to supervisors and managers. Pursuing low LTIs as a safety measure brings a form of 'manufactured insecurity' (Taylor, 2023). People can be watched at any time. And the consequences of a 'bad' LTI number can be felt any time, too – wherever you are in the organisation.

The modern word 'insecurity' entered the English lexicon in the 17th century, just as a market-driven society began. It is interesting to reflect on how attempts to create greater security in the following centuries – physical, medical, food, hygiene, and more – have an inevitable shadow side created by the exact same mechanisms. The insecurities introduced by pursuing security through rationality, planning, measurement, and science have been captured by what Beck (1992) called a risk society. A risk society needs to learn to contend with the new and not always entirely predictable human-made insecurities that come from its attempts to harness and control the world's complexity while squeezing productive gain out of it (cf., Giddens, 1991). The desire for security and predictability in an inherently uncertain world (made uncertain in part by our incessant attempts to make it more secure) often leads to profound dissatisfaction and unease (Watts, 1951). The pursuit of security manufactures new insecurities.

Since the 1980s, the neoliberal turn has produced market-favouring policies across the Western world, increasing the stakes of manufactured insecurity. The suspicion, if not an accusation, was that workers employed by state-owned enterprises were cogs in bureaucratic machines with no space for initiative or free will. In contrast, in an unhampered capitalist system, they would be free people whose security and liberty would be simultaneously guaranteed by an economic democracy where everything could be had, traded, sold, or bought for a price that the market determined. Of course, it didn't turn out that way. Workers in non-state enterprises are now routinely subjected to a panoptic regime of surveillance, management, quantification, measurement, monitoring, assessment, and control.

The effects on people's autonomy and sense of professionalism are corrosive, Lorenz (2012) concludes. Lorenz points out that we have gradually adopted an Orwellian language that redefines quality, accountability, responsibility, professionalism, and even freedom. This language tends to invert, if not pervert, these terms into the opposite of their original meanings. Quality means compliance – not straying outside a narrow bandwidth of approved performance. Accountable means compliant – you can show or be accountable for how you followed the rules. Responsible means compliant – taking ownership for nothing more or less than seeing that compliance expectations are met. Professional means compliant – somebody is a professional when they don't flout the rules. Freedom comes to mean compliant – not with what the government wants you to do, but what a free market demands you to do so that you can settle the accounts for your successes and failures (or injuries). Indeed, a part of the surveillance and the language around it is directed at the prevention – or at the possibility of renaming or denying – injuries and incidents so they don't muddle the quarterly books. The specific forms in which the contradictions of market freedoms have manifested themselves may have been hard to predict or expect. But the promised freedom turns out to be freedom for capital, gained at the expense of human security.

The paradox has long been baked into modern Western society. Philosopher Jeremy Bentham wrote in 1802 that when our security reaches a certain point, the fear of losing it prevents us from enjoying it. As he puts it, the care of preserving condemns us to a thousand sad and painful precautions, which are always liable to fail. Organisations under pressure to give in to Saul's 'Quarterly Syndrome' have a thousand sad and painful precautions on their books and throughout their workplaces – including the endless clutter to prevent, rename, or manage injuries and incidents so they don't muddle the books. However, these sad and painful precautions are always liable to fail. It is the consequence of living in an insecure and risk-filled world, despite (or partly because of) our being pumped with the illusion that we've got it under control – and a single digit to show for it. We might proclaim (and, indeed, honestly believe) that safety is our top priority. But the manufactured insecurity of LTIs reflects a more cynical take on the human condition: we will work well (or show the numbers that suggest we work well) only under the threat of loss and duress. Constant insecurity helps keep all of us in line. But what for? And isn't it working against all of us?

2

STOP COUNTING WHAT YOU CAN COUNT – IT DOESN'T COUNT

For years, businesses worldwide have relied on one measure, the total recordable incident rate (TRIR), to gauge safety performance. Some countries use the total recordable injury frequency rate (TRIFR), a variant of TRIR tailored to local conditions and regulations. These measures, pivotal in shaping board and business decisions and share valuations, have become the de facto yardstick by which to measure the success of an organisation's safety management.

At first glance, this might seem perfectly reasonable. These are, after all, internationally recognised standards that industries have relied on for decades. They quantify safety incidents, providing a seemingly concrete, comparable figure on which to pin safety performance. This number influences everything from business contracts to the daily routines of workers on-site. But what if our trust in these metrics is misplaced? What if they are not the stalwart sentinels of safety that we perceive them to be?

DOI: 10.4324/9781003177845-2

Let's dig deeper into scrutinising TRIR and its variant, TRIFR. We need to question their efficacy as decision-making tools. It is not an effort to dismantle tried and tested systems. Instead, it's an invitation to re-evaluate, investigate, and perhaps look beyond what we've come to accept as an industry norm.

What is TRIR?

In workplace health and safety, statistical analysis is crucial in understanding and evaluating various variables. It does so for the fundamental metric used to assess safety performance as well, the total recordable incident rate (TRIR). TRIR serves as a vital indicator, offering insights into the prevalence of injuries and illnesses that conform to OSHA's recordability standards within a given organisation.

TRIR, being a rate, comprises two essential components: the number of recorded events and the corresponding time during which these incidents occurred. In this context, the events refer to work-related injuries and illnesses, whereas time is measured in worker exposure hours. Combining these elements, TRIR quantifies the frequency of recordable incidents relative to worker hours, providing a meaningful representation of safety performance.

To avoid dealing with tiny fractions, TRIR is scaled per 200,000 worker hours, a value that approximates the working hours of around 100 full-time employees over a year. Consequently, the TRIR also conveys the percentage of workers who experienced a recordable incident over a year. This standardised approach allows organisations to assess and compare safety performance across different timeframes and workforce sizes.

Equation 1:

$$TRIR = \frac{\text{Number of Recordable Incidents} \times 200{,}000}{\text{Number of Worker Hours}}$$

Organisations calculate the total recordable incident rate using Equation 1 for specific periods. For instance, if a company recorded three reportable incidents during a month, and the total worker hours amounted to 350,000, the TRIR would be computed as 1.71 per 200,000 worker hours.

Typically, TRIR is reported as a single numerical value, assumed to represent safety performance for that specific period as the only possible outcome. Moreover, subtle differences in TRIR, even to decimal points, are often considered meaningful in tracking safety trends over time.

What is TRIFR?

The total recordable injury frequency rate (TRIFR) is a safety performance metric that is similar to TRIR (total recordable incident rate) but is more commonly used in some regions, especially in specific industries (e.g. the mining industry in Australia relies almost entirely on TRIFR – and what the mining industry does, most Australian industries do). TRIFR is explicitly focused on measuring the frequency of total recordable injuries within a defined population of workers during a specific period. It is an essential tool for organisations to assess and monitor their safety performance and identify trends in workplace injuries.

The history of TRIFR is closely tied to the broader evolution of safety management practices and the supposed need for standardised and meaningful metrics to evaluate workplace safety. It emerged as a refinement of traditional incident rate calculations, aiming to provide a more comprehensive view of the safety performance of an organisation or a specific industry.

The formula for calculating TRIFR is relatively straightforward:

TRIFR = (Total Recordable Injuries/Total Hours Worked) × 1,000,000

Here's a breakdown of the components:

Total Recordable Injuries: This includes all work-related injuries and illnesses that meet the criteria for recordability as defined by the relevant regulator. These incidents typically involve medical treatment beyond first aid, restricted work activity, or lost workdays.

Total Hours Worked: This refers to the sum of all hours worked by the employees during the defined period. It includes both regular working hours and any overtime or additional hours worked.

1,000,000: The multiplication by 1,000,000 is a scaling factor used to express the rate per one million hours worked. This makes the metric more manageable and more accessible to interpret.

By calculating TRIFR, organisations can quantify the frequency of recordable injuries and illnesses within their workforce on a standardised basis, making it easier to compare safety performance across different periods and companies. Lower TRIFR values have been seen as indicating better safety performance and a safer work environment.

Understanding the Construction Safety Research Alliance (CSRA) study

In 2020, Hallowell and colleagues (Hallowell et al., 2020) decided to join forces with executives from ten different construction companies to conduct an in-depth study focusing on safety in their industry. They had a massive dataset of over three trillion worker hours, including all reported incidents from the participating companies. Was TRIR telling them something authentic? Something useful? Their analysis was parametric as well as non-parametric.

Parametric analysis is like assuming all apples in a basket are identical. This method allows the researchers to speculate that the data follows a specific pattern, such as a binomial distribution (a fancy term for a particular type of pattern in statistical data). This assumption, in turn, helps the researchers evaluate the accuracy of TRIR by calculating what's known as confidence intervals, which are a bit like the margin of error on a political poll. Confidence intervals provide a range of values within which the actual TRIR is likely to lie. In a nutshell, parametric analysis aids in determining how precise and reliable TRIR is as a safety measure.

On the other hand, non-parametric analysis doesn't assume all apples are the same. Instead of assuming a specific pattern in the data, it estimates the distribution based solely on the available data, like judging a group of individuals based solely on their unique characteristics and not on any preconceived notion. This type of analysis is instrumental in testing whether past TRIR values can foretell future TRIR values. In essence, it helps researchers see if there's a correlation between a company's past safety track record and its future safety performance.

By leveraging these analytical methods, the researchers hoped to better understand when and how TRIR can be used to compare different companies or predict future safety performance. Data-driven techniques were the key to dissecting and understanding the extensive amount of incident data. By amalgamating the parametric and non-parametric analyses, the researchers were able to furnish more robust insights into the reliability and predictability of TRIR as a safety performance measure.

Three key assumptions

Let's picture a busy construction site – workers in hard hats and safety vests hustle and bustle around, carrying materials, operating heavy machinery, and shouting orders. Amidst all this activity, accidents can sometimes happen – a misplaced foot can lead to a tumble, or a momentary lapse in concentration might cause a minor injury.

We want to analyse these incidents that occur during a worker's shift. Imagine each working hour as a separate event or TV show episode. In this context, an incident can happen during that specific working hour (like a cliffhanger moment in the episode), or it may not happen (an uneventful episode).

Here are the three core assumptions we work with:

Binary outcomes: The first assumption is like a light switch: either on or off, and there's no in between. In other words, during each working hour, an incident either happens (yes) or it doesn't (no). There's no room for partial incidents; it's like saying someone is 'a little pregnant' – it's a yes or a no. So, in each working hour, an accident occurs, or it doesn't: as simple as that.

Constant Probability: The second assumption is that the chances of an incident occurring during a working hour remain relatively consistent, like the probability of rolling a six on a die. Sure, some situations might pose a higher risk (just as you might have more sixes in a hot streak during a board game). However, over many working hours, the overall probability balances out and becomes relatively consistent.

Independence: The third assumption is about independence; it's like each episode of the TV show stands alone and isn't influenced by what happened in the previous episodes. In our case, the outcome of one working hour doesn't affect what happens in the next. It's like each working hour is its own bubble. Even if an accident happens during one hour, it doesn't make an accident in the next hour more or less likely. We make this assumption because, when dealing with a large dataset with countless working hours, any potential patterns or connections in the occurrences of incidents get diluted and become less significant.

Confidence intervals: The 'safe' range

Let's tackle this with an everyday example. Picture going to a car mechanic to fix an engine issue. He could say, 'I'll have it fixed in three hours.' But, experienced as he is, he knows very well that unexpected issues can pop up, or things might go more smoothly than anticipated. So, instead, he gives you a range: 'I'll have it done in two to four hours.' This, in essence, is like a confidence interval: a range within which the actual outcome is likely to fall.

Let's relate this to our topic, the total recordable incident rate (TRIR). Suppose we had a large construction site bustling with activity. Every hour a worker is on-site, we can ask a 'yes' or 'no' question: Did an accident happen in this hour?

Picture these answers as different coloured balls (green for 'no' and red for 'yes') in a large box. The distribution of these coloured balls in the box represents a Poisson distribution. This distribution helps us calculate the likelihood of specific injuries occurring in a set number of working hours, like guessing how many red balls might be in a particular section of the box.

In a span of 200,000 worker hours, a company recorded 1 injury. To understand this better, we calculate a confidence interval: a 'safe' range in which we believe the true TRIR will likely be found. Using the Poisson distribution and a nifty mathematical tool called the Wilson confidence interval equation, we figure that the TRIR could range from 0.18 to 5.66 injuries per 200,000 worker hours.

Think of it like the mechanic's time estimation; instead, it's a range of possible injury rates. Reporting a single TRIR number, like 1.00 per 200,000 worker hours, is like the mechanic giving you an exact time of three hours – it lacks flexibility. It doesn't account for the unpredictability of real-life situations. Instead, giving a range (0.18 to 5.66), with the most probable value of 1.00, is more informative and accurate.

The beauty of the confidence interval is that it also helps us speculate about the future. Say the company's actual injury rate is 1 per 200,000 worker hours. We can use the confidence interval to guess how many injuries might occur in the future, provided the safety conditions remain the same. It's a bit like a weather forecast, giving us a range of possible outcomes and their chances.

For instance, if there are 2 injuries in the next 200,000 worker hours, our new confidence interval would be between 0.55 and 7.29 injuries per 200,000 worker hours. Even though the injuries have doubled from the last period, a statistical test indicates that this difference is insignificant. This suggests that the number of injuries alone doesn't reveal anything crucial about the changes in the safety system between the two periods, much like having two rainy days in a row doesn't necessarily mean the whole month will be rainy.

Understanding TRIR:
Unpacking three company case studies

The study (Hallowell et al., 2020) then delved further into the safety performance world by examining three fictional companies: A, B, and C. They explored how the concept of the total recordable incident rate (TRIR) as a confidence interval can alter our understanding of each company's safety situation.

First up: Company A. The company was new and had just started up, and within their first 1,000 worker hours, they had 1 unfortunate incident. If their TRIR was reported as 200 (per 200,000 worker hours), it might sound like they're running a risky operation. But wait, let's bring in the handy confidence interval. When that's done, it's discovered that Company A's TRIR could range from 35 to 1,128 incidents per 200,000 worker hours. That's a broad range, revealing that understanding their safety performance isn't precise. Since Company A is new and has fewer worker hours, the range is wide, meaning the exact TRIR could be anywhere in that range.

Moving on to Company B, which had 7 incidents over 980,000 worker hours over a year. A straight TRIR of 0.69 (per 200,000 worker hours) might make them seem like a low-risk company. However, don't settle for appearances, right? When the confidence interval is applied, the TRIR range becomes 0.69 to 2.95. Even though they've performed better than Company A, there's still a significant range in which the real TRIR could fall.

Lastly, let's look at Company C. They had a busier year, with a whopping 6,000,000 worker hours and 24 incidents. Reporting only the TRIR, 0.54 (per 200,000 worker hours) might make it seem like they're on par with Company B. But hold on – the confidence interval tells a different story. It narrows the TRIR range to 0.54 to 1.19, a more precise range than the others.

So what can be learned from these examples? The number of worker hours plays a massive role in understanding TRIR. The more worker hours, the more precise TRIR values become. But here's a word of caution – no matter how many worker hours, don't be tempted to report TRIR to a high level of decimal points (1.00 or 0.95) without enough data.

For instance, to report TRIR to one decimal place (like 1.0), a company needs about 300 million worker hours. And for two decimal places (like 1.00), they'd need a staggering 30 billion worker hours. With injuries, thankfully, being rare events, getting these precise TRIR values is a tall order.

Even if an organisation reports a TRIR of 1.0 over 1,000,000 worker hours, it doesn't mean the actual TRIR is precisely 1.0. It indicates that their 'safety system' will likely produce a TRIR between 0.43 and 2.34 per 200,000 worker hours 95% of the time.

Crunching the numbers: The non-parametric analysis

Let's go for a deep dive into another type of data analysis called the non-parametric approach, which the CSRA also used in their study (Hallowell et al., 2020). Think of it like taking a second look at the crime scene in a detective novel. This time, they wanted to confirm what they found earlier and maybe uncover some additional clues.

To do this, they gathered real-life data from ten construction companies in a big group called the Construction Safety Research Alliance (CSRA). These aren't your regular small-scale companies. We're talking about heavy hitters in infrastructure, power generation and delivery, and commercial building projects.

These companies opened their books and provided data for the number of recordable injuries, fatalities, and worker hours they clocked each month. And they didn't just give data for a year or 2; they provided an extensive dataset spanning 15 years. If you're wondering how big that is, it's a colossal

3.26 trillion worker hours' worth of data. That vast pool of information provides a broad and deep look into the construction industry.

So what did the CSRA do with all that data? They put it under the microscope in an empirical analysis. An empirical analysis is a bit like assembling a massive jigsaw puzzle. You study each piece (or data point) and find where it fits in the bigger picture. Doing this with such a vast amount of data helped confirm the earlier findings from the parametric analysis. Plus, it gave additional insights into patterns and trends related to workplace injuries and safety performance in the construction industry.

If you're a data geek and want to know more about this empirical analysis, there's a research paper by Salas (2020) in which you can find all the details. It explains the methodology and results found, giving a complete account of this massive data detective work.

Decoding TRIR

When figuring out the value of the total recordable incident rate (TRIR), the CSRA used a generalised linear model (GLM) to test out different distribution functions. Think of it like trying on different pairs of glasses to find out which one helps you see the best. The idea was to find the best model to make sense of the data from different companies.

Then they tested this model like a car in a test drive. They used the historical TRIR data like a roadmap to see if they could predict future TRIR values, like predicting where the car would go. The results were a shocker. The model could only predict the TRIR in about 2% to 4% of the cases. So TRIR turned out to be like a wild horse: highly unpredictable.

When you think about it, it does make sense. Workplace safety isn't a simple tic-tac-toe game; it's more like 4-D chess. It involves many elements like the workers, company culture, regulations, machinery, and external things like the economy, weather, and other natural events. With all these factors stirring the pot, it's no surprise that TRIR is hard to predict accurately.

The randomness of TRIR teaches some vital lessons. First, it backs up the logical assumptions made in the earlier analysis. Second, it highlights that TRIR should be shown as a range (a confidence interval), not just a single number. This way, you get a broader and more real-world view of what could happen with safety performance.

The researchers also tested how well the past TRIR data could foretell future TRIR values. They found they needed at least 100 months of data to have a decent shot at prediction because TRIR varies so randomly. Since companies often use TRIR to make decisions over months or years, this means that TRIR isn't a crystal ball for predicting future safety performance.

In real terms, imagine a company hiring a contractor with a TRIR of 0.75. Expecting the contractor to score the same on their next project is unfair. TRIR is a bit like a weather forecast; it can vary unpredictably, making it a tough cookie to use as a reliable predictor for future safety outcomes.

Busting the myth of a connection between TRIR and fatalities

You are wrong if you think a lower total recordable incident rate (TRIR) means fewer workplace fatalities. The numbers crunchers have done their homework, and the results were clear – there's no connection between them. In other words, the TRIR changes don't tell us anything about the likelihood of fatal accidents. It's a bit like eating more ice cream in the summer doesn't cause more sunburns. The two happen to occur at the same time but aren't connected. Instead, fatal incidents seem to dance to their own tune and have unique patterns. It's like they happen for entirely different reasons. This discovery shakes up some old beliefs, specifically those from the Heinrich safety pyramid (Bird & Germain, 1985; Heinrich et al., 1980).

It's like saying that for every 100 paper cuts, there would be ten serious wounds and one fatal injury. His idea was that if you can reduce minor injuries, you'll also see fewer severe and fatal injuries. But the CSRA's findings and many others have busted this myth.

The new data suggests that reducing TRIR isn't a magic bullet for avoiding serious accidents. It's like brushing your teeth to prevent hair loss – the two aren't connected. So we must look at other ways to build resilience and miti-gate the risk of such high-impact workplace events.

The findings in a nutshell

Most safety professionals have been trained to think that TRIR is a solid benchmark of safety performance. But what if everything we thought we knew was wrong? This study took a deep dive into a whopping 3.26 trillion worker hours of TRIR data, and what was found challenges everything.

Firstly, TRIR doesn't predict fatalities. Getting a paper cut at work (a recordable injury) and having a fatal accident are two different things. The CSRA research showed that the two don't follow the same patterns or happen for the same reasons. So focusing on lowering TRIR won't necessarily prevent fatal accidents. It means we have to rethink our safety interventions – the policies, regulations, and management systems that aim to improve TRIR might not do much to prevent fatal incidents.

Then there's the wild randomness of TRIR. With 96% to 98% of changes in TRIR happening due to sheer randomness, it's like a lottery. It makes sense when you consider how recordable injuries don't happen in predictable patterns or at regular intervals. Safety is complex and influenced by many factors (Dekker et al., 2011).

Because TRIR is so random, it's misleading to represent it as a single number. It's like judging the quality of a TV show based on a single episode – it doesn't give you the whole picture. A better approach is to express TRIR as a range (a confidence interval) and observe it over extended periods. For instance, a yearly TRIR of 1.29 doesn't mean much for most organisations. And the level of randomness in TRIR means it's not practical or accurate to report it to several decimal points.

Using TRIR for performance evaluations can be problematic. Since TRIR is mainly random, it's hard to tell if a change in performance is due to an actual change in the safety system or just a twist of fate. Rewarding or penalising based on TRIR might mean rewarding or punishing randomness.

TRIR can predict future performance, but only over very long timeframes. Some past studies suggested that TRIR could be a helpful predictor if taken over long periods (Alruqi & Hallowell, 2019; Lingard et al., 2017; Salas & Hallowell, 2016). The CSRA research confirms this, but you'd need over 100 months of TRIR data to see any predictive value.

What are the implications for safety management?

Stop using TRIR as a stand-in for severe injuries and fatalities: TRIR has long been seen as the oracle of safety trends, with a downward TRIR trajectory interpreted as reduced fatal accident risk. But here's the thing: Our research found no statistical link between TRIR and fatalities. So this belief isn't scientifically grounded. We must design and test focused measures and preventive efforts for severe injuries and fatalities.

Don't use TRIR to benchmark performance or draw comparisons: TRIR isn't the reliable yardstick we thought it was. To get a reasonably precise TRIR figure, you'd need tens of millions of worker hours. Most companies can't even generate enough worker hours to detect significant year-to-year TRIR changes. So comparing companies, projects, or teams based on TRIR isn't a fair game. At most, TRIR might help compare broad sectors of the US economy over extended periods. Therefore, we should rethink using TRIR as a critical safety metric in performance incentives or contractor prequalification.

Change how we communicate TRIR: TRIR is typically touted as a single, accurate number as though it's gospel truth. But given how rare recordable injuries are and the high degree of randomness involved, this number isn't as significant as we've made it out to be. Instead, we should present TRIR along with the range of possible outcomes that our safety systems could reasonably produce. In this way, smaller companies would report a broader range (reflecting higher uncertainty), whereas larger companies would report a narrower range (showing lower uncertainty). However, almost no company could report TRIR with the level of precision commonly used today.

Embracing these insights can lead us towards more meaningful, adequate safety measures that can make a real difference in protecting workers.

3

HOW MANY INJURIES ARE ENOUGH?

People commonly judge how safe a workplace is by looking at the number of reported injuries. But there are some concerns.

Firstly, when a workplace reports only a few injuries, those numbers don't offer enough data to make meaningful comparisons over time. As a workplace gets safer, these low numbers of injuries become even less helpful. Fewer injuries mean less data, and it's harder to say whether safety improvements are genuinely making a difference or if it's just luck.

Secondly, the typical injury stats often show more about minor accidents – like tripping over a loose carpet – than severe or fatal incidents. You may think your site is safe because it only had a handful of 'Oops, I tripped' reports last year, but you are probably missing the bigger picture.

Reported injuries can say a lot about workplace culture and the ability to be honest about them in the first place (Dekker, 2023; Edmondson, 1999; Saloniemi & Oksanen, 1998; Sherratt & Dainty, 2017). In this sense, a site with fewer injuries is probably a much less safe place – in both psychosocial and physical ways.

DOI: 10.4324/9781003177845-3

Lastly, it turns out that the number of reported injuries can be more a reflection of how the organisation reports or categorises these incidents than of the actual safety level of the workplace. Different places have different reporting habits, which can muddy the waters when comparing safety records.

The selective reporting study

A study by Geddert et al. (2021) zeroed in on that last issue. They wanted to know if the usual way of measuring safety, often called lost time injury frequency rate (LTIFR) data, is more about bureaucratic procedures than fundamental, on-the-ground safety. They created a new method that compared reported incidents to the actual outcomes. Specifically, they wanted to figure out if incidents marked as 'recordable Injuries' really do reflect the most significant types of injuries that happen at a workplace.

Previous studies have shown that not all injuries get reported. For example, a survey of Canadian workers found that 40% of those injured didn't even file a claim. And get this: another study concluded that although 30% of workers experienced conditions that should have been recorded, less than 5% reported them.

Other research looked at different sources of injury data and found significant discrepancies. One study noted that employers didn't report 80% of construction workers' claims to the safety authorities (De silva et al., 2017). This demonstrates that our data isn't necessarily telling the whole story.

The research by Geddert et al. conducted 'one-to-one injury matching.' They compared the same injury report from two sources – like a company's internal records and external data from WorkCover. This gave two perspectives on each incident and helped create an understanding of what kinds of injuries get under-reported or misclassified.

There were four key questions to answer:

1. What proportion of injuries that require medical treatment are officially recorded?
2. How does the severity of an injury relate to whether it's recorded?
3. Can we see any patterns between the risk classification of an injury, its actual severity, and its recording status?
4. Are there any trends between the type of injury (like what body part is affected) and whether it gets recorded?

The research aimed to determine if the way injuries are recorded affects a company's understanding of their workplace's risk. It is important to note that the researchers tried this approach at an energy company in Australia, where workplace injuries were generally covered by state-run insurance, with a few exceptions. They got data from the energy company and state insurers for a balanced view.

The energy company used a so-called incident management system (IMS): proprietary software to track all adverse (or potentially adverse) events, such as injuries or near misses. They also rated these incidents based on their risks, a requirement of Australian law. As a result, they gleaned various types of data from this system, such as the kinds of incidents that were happening, which body parts got injured the most, and so on.

The state insurer also shared some data with the researchers, such as the type of claim made, how much was paid, and the legal costs involved.

To connect the dots, the researchers also ensured that records from both sources could be matched. They studied all the claims from 2018 and 2019. Each injury got its unique record; data was combined from both systems. For example, if there was a disagreement about which body part was injured, the trusted data was the insurer's data. The researchers also simplified some categories, like combining all hand-related injuries into one. Sometimes, entries in the IMS system didn't match the insurer records. In those cases, they were flagged as 'not work related,' which kept them consistent with how the company usually handled such discrepancies. This brought out some interesting insights, such as how some incidents classified as 'not work related' by the company are considered 'work related' by the insurer.

The severity of each injury was then classified based on a few criteria, including the cost associated with it. This helped divide the cases into four bands, ranging from 'low' to 'very high' severity. But it's important to remember that costs can vary due to several factors, like the employee's salary or whether the company was found to be negligent.

The study (Geddert et al., 2021) was based on one company's data. The company operated onshore oil and gas rigs with their own set of risks. They were seen as thorough about safety, using a particular analysis technique called systematic cause analysis technique (SCAT) and making risk assessments right when an incident occurs. The researchers did want to bring in more players, but companies can be a bit protective about this kind of information. A total of 53 incidents were looked at, all involving some form of medical treatment deemed work related by the insurance company.

This study showed that the way work incidents are reported might not be accurate. Only 19% of insurance claims were initially marked as 'recordable,' which means we might be underestimating some incidents' seriousness. This could call into question the belief that under-reporting is mostly because workers don't tell their companies about incidents. Using this method also had other benefits:

- First, using the one-to-one method means we can tell if more severe injuries were more likely to get reported. For instance, in the test company, severe injuries were more likely to be logged than minor ones (25% compared to 15%).
- The method can also show if there's any under-reporting going on, even for serious injuries. The data showed that only 17% of the bad injuries and 33% of the really bad ones were listed as 'need to record.'
- If we want to know if the 'recordable' injuries are the most severe, this method can help us measure that, too. Surprisingly, the chances of a 'recordable' injury being minor or severe in the test company were about the same.
- It also helps us understand why some severe injuries weren't marked 'recordable.' Often, they were labelled 'not work related.' Out of 12 major injuries, only 3 were tagged as 'recordable.'
- The method also reveals that the risk ratings assigned during the accident weren't in sync with the actual outcomes. The agreement level was super low at 1%.
- Back and hand injuries were more common and severe in the test company. This method let the researchers compare how often these body parts were injured and how often those injuries were considered 'recordable.'
- Biases were highlighted in the types of injuries classified as 'recordable.' For example, most pinch and crush injuries were recorded, but only a tiny percentage of injuries caused by exertion made it into the record.
- Also noticed was that the severity of hand injuries correlated with them being recorded. But oddly, the more severe the back injury, the less likely it was to be recorded.

Key findings

1. Through the study, it was identified that not all injuries were recorded correctly. The study found that only 19% of the insurance claims were labelled 'recordable injuries.' However, these injuries were still work related and often required medical treatment. The reasons for the mismatches were:

   • Sometimes injuries needed more than first aid but were only recorded as 'first aid.'
   • Some injuries were considered 'not work related,' even when insurers later said they were. You might think companies are trying to 'hide' these injuries, but it seems like a more significant issue that's common across industries.

2. High- and very high-severity injuries were more often labelled as recordable, but many still slipped through the cracks. It was also found that the total cost of injuries labelled 'not work related' was even higher than that of the recorded ones!
3. There is no real connection between risk ratings and actual injuries. This means the company is not just under-reporting injuries but is also misunderstanding where the real risks lie.
4. Some injuries were more likely to be under-reported than others. For instance, hand injuries were often recorded, but back injuries? Not so much. This suggests that companies might have a blind spot regarding specific injuries.

Accurate injury reports are the bedrock of a company's safety plan. If those are off, everything else will be, too. It's commonly known that there's a lot of pressure across various industries to keep the number of 'recordable injuries' low, maybe because companies use this metric as a performance measure. Think about it: We pay attention to what gets reported, so if only specific injuries get the spotlight, we might miss other serious threats. That's like being in an echo chamber, where you only hear what you want to.

How we classify 'recordable injuries' is a big deal in many industries. If we're not doing that right, we need to own up and fix it. The study hints that our current systems might prevent us from recognising specific important injuries, and our definitions are skewed. The industry might define what's risky based on 'recordable injuries' rather than on what's happening.

Not having a clear idea of what's risky can skew our understanding of safety. For instance, less severe injuries might get brushed off as 'first aid' or 'not work related' instead of being used as learning opportunities. Instead of guessing how severe an injury is, we should consider costs like medical bills and time off work. This can help us get a more accurate picture and better manage safety in the long run.

Using lost-time injuries

Some companies measure 'safety' using lost-time injuries (LTI) rather than just TRI. As it turns out, that's also a shaky way to judge how safe a workplace is. Dr Marloes Nitert and Dr Sidney Dekker worked with a company that was super proud of lowering their injury rates over four years, going from 19 injuries to just 1. Management was over the moon, thinking their safety posters and extra training sessions were working miracles. But here's the twist: the numbers were so small that it's tough to say if their actions made a difference. In simpler terms, their sample size was so small that the change in injury rates could just have been random luck, not the result of their safety programs. When the numbers were crunched, they found a 92% chance that it was all just random noise.

To be sure something isn't just a random fluke, researchers like a 95% confidence level. To reach that confidence level with their tiny staff of 85, the company would need to have initially recorded 20,400 injuries and brought it down to 1,020. Either that, or they'd need to employ about 53,000 more people to make their numbers reliable. 'Statistically significant' means there is a way to be confident that a change (like fewer injuries) didn't just happen by dumb luck. It's a way to say, 'Hey, what we're doing is making a difference.'

Let's bring it back to that company with 85 workers. They worked three 8-hour shifts, five days a week, totalling 170,000 hours a year. Over four years, the injury rate dropped from about 0.011% to 0.0006%. This looks great on paper; however, when you do the math, you realise they likely just got lucky. Even if the manager could show a significant drop in injuries, it's not necessarily because of what they did. It could just be good luck this year and bad luck the next. If we base our safety programs on these shaky numbers, we might think we're doing great when we're not. Or, even worse, we could be reinforcing some terrible ideas.

In summary, small numbers can trick us into thinking we're doing better (or worse) than we are. It's a kind of 'tyranny of small numbers,' in which a small sample size can make things look more significant than they are, so the next time someone brags about how their safety initiatives lowered injuries, it should be taken with a grain of salt.

4

THE ABSENCE OF INJURIES DOES NOT PREDICT WHAT YOU THINK IT DOES

Since the 1940s, fatal workplace accidents have decreased, which is good news, but specific industries like mining, offshore work, and construction still see more of these tragic incidents. Studies have shown that fatal accidents mostly happen to men on the lower rungs of the job ladder. Race and being a migrant worker also seem to play a role. These fatal accidents can sometimes indicate more significant workplace equality and safety issues. Interestingly, some studies suggest that workplace accidents might increase when the economy is booming. It's like a double-edged sword: good times for businesses but possibly riskier for workers.

In an article by Saloniemi Oksanen, data from Finland focusing on the construction and manufacturing industries from 1977 to 1991 found surprising results. While some indicators like working hours and unemployment rates didn't strongly relate to the rate of fatal accidents, other factors did. In construction, fatal accidents increased when the number of general

DOI: 10.4324/9781003177845-4

accidents decreased. The study used statistical tests to understand how various factors, like the number of employees and working hours, affect accident rates. Remember that these stats are based on recorded accidents, so the data could have limitations. We know that understanding the reasons behind workplace accidents isn't straightforward. Different factors like industry type, economic conditions, and even the type of accident can unexpectedly influence the stats. It's a complex issue that needs more investigation.

The problem with 'safe' systems

As far back as 1978, experts suggested that what we do to keep safe might unintentionally create new risks. More recent incidents have brought this idea back into the spotlight. It turns out that the standard ways of keeping things safe aren't as effective as we thought. Things like safety orientations, keeping records, and even emergency planning don't necessarily improve safety that much. On the other hand, things like strong leadership, good sub-contractor management, and employee involvement in safety seem to work better. Some experts say we still hold on to 'safety myths' like 'more barriers mean more safety' or 'human error is always the root cause of disasters.' These myths are sometimes the reason progress in safety gets stuck.

Following on from that, bureaucracy can create more problems than it solves. Sometimes, the systems designed to spot and count risks hide those risks. This doesn't just happen because there's too much paperwork, although that's part of it. Sometimes, companies have years without incidents and get lulled into a false sense of security. When warnings appear, they're often ignored because 'That can't happen here.'

In an article by Dekker and Pitzer (2016), research was reviewed to see if our current safety practices might unintentionally worsen things. It looked at various assumptions like 'more rules mean more safety' and asked whether these might be doing more harm than good. The idea that minor incidents predict significant accidents is not supported by evidence. For example, a construction site with fewer minor incidents might have a higher rate of serious accidents. There's also a tendency to think, after the fact, that we could have seen an accident coming. This leads to overconfidence in our safety measures.

In the early 1900s, so-called 'Taylorism' suggested that bosses and engineers were the smart ones who made all the rules. Workers should follow these rules to the letter for things to go smoothly and safely. It's as if the people who wrote the rule book had it all figured out, and everyone else should stick to the script. But work is messy, and there's often a difference between what the rule book says and what happens on the ground. Sometimes, people have to make quick decisions or adapt to situations. So saying that someone is 'not following the rules' can be a pretty harsh judgment that overlooks how creative and resilient workers often have to be. Recent studies show that some best practices come from people's 'on-the-ground' adjustments. Rules and procedures are helpful but can't cover every possible situation. Sometimes, you need skilled people who can adapt and think on their feet to keep things safe.

There's also the issue of how we think about risk and control in complex systems. As things get more complicated, it's harder to pinpoint what's risky and what isn't. And honestly, no one person or even department can control all the risks. So when companies rely only on paperwork or audits to say, 'We've controlled the risk,' that can sometimes be wishful thinking. It's more about covering their backs than actually making things safer.

This, of course, brings us back to the idea of 'human error.' For a long time, safety efforts have focused on stopping individual workers from making mistakes. But that's missing the point. It's not just about individuals goofing up; it's about the whole system and the conditions people work under. If you want to improve safety, you can't blame the workers and try to fix their behaviour. You have to look at the big picture. Instead of pointing fingers or doubling down on rules, we should appreciate that people, especially those dealing with the nitty-gritty, usually do their best to adapt and make things work. Their expertise can be a crucial part of making workplaces safer.

Many companies aim for 'zero harm' or 'zero incidents' in their workplaces. It sounds great and may supposedly meet the strategic goals of the C-suite. But sometimes, the way companies track safety numbers can make workers less likely to report when something goes wrong. One reason is that nobody wants to be the person who ruins the 'zero accidents' record. A US government study (GAO, 2012) investigated this and found two types of safety programs. One focuses on having low rates of reported injuries, and the other rewards workers for doing specific safe behaviours. The first kind can discourage people from reporting injuries because they don't want to mess up the stats. Other policies, like drug tests after an accident, can make workers hesitant to speak up.

The study also pointed out that creating an open culture in which people feel free to discuss safety issues can help. But just focusing on numbers can hide the real story. And when bonuses and promotions are tied to these numbers, people can manipulate the data or keep quiet about issues. This can lead to overlooking serious safety concerns. Suppose a company is obsessed with keeping injury numbers low. In that case, it might not be paying attention to more significant safety issues that haven't caused an accident yet but could in the future.

The under-reporting also ties in with Heinrich's iceberg model, which says that for every significant injury at work, there are lots of smaller incidents leading up to it – like minor injuries and near misses. For example, in Finland, for every 12 major injuries that keep someone out of work for over a month, 101 minor injuries cause at least three days of absence. But the theory is not without controversy. Heinrich believed that the same underlying issues cause both minor and significant injuries. Some studies, like one that looked at over 7,000 injuries in the oil industry, back this up. They found that what people did before a minor injury wasn't that different from what they did before a major one.

However, not everyone agrees. Another researcher, Petersen, argues that different situations give rise to minor and major injuries. He points out that some types of incidents – like transportation accidents – are more likely to cause severe or fatal injuries, whereas others, like slips and falls, might result in minor injuries. This debate matters because if we're trying to make workplaces safer, we must know what we're dealing with. Are all injuries stemming from the same contributory factors? Or do we need different strategies for minor and major injuries?

Work by Dekker and Pitzer (2016) aimed to answer those questions by looking at real-world data rather than just theorising. The authors examined research findings from a study completed by Salminen et al. (1992) grounded in real-world incidents. They focused on serious workplace accidents in Southern Finland from September 1988 to December 1989. First, they looked at 99 severe accidents in Southern Finland over about a year. In these incidents, 102 people got hurt, and, sadly, 20 of them passed away.

All the fatal cases were men, and 12 injured were women. By 'serious accidents,' the researchers meant anything from losing a fingertip to, unfortunately, fatal incidents. Even situations in which a severe accident was possible (for example, falling from a height) were included. They left out minor accidents, such as getting something in your eye. To figure out what happened, the researchers visited the accident sites and talked to people who saw them happen. They used a particular method to dig into the details. In line with a Finnish research model, they mapped out how the person and the hazard collided, leading to the accident. There was a lot of data to analyse – like what factors led to the accident. This could have been anything from a poorly designed workspace to risky behaviours by workers. These factors were then divided into 14 categories, including things related to the victims, the workplace, equipment, etc. The agreement between two people independently categorising these factors was about 78%.

The findings of the study were fascinating. The kinds of accidents that led to deaths differed from those that were not fatal. Specifically, people were more likely to die if hit by a falling object or run over by a moving vehicle. Falling from a height or machine-related accidents were less often fatal. Across 99 accidents, 1,189 factors were identified, averaging about 12 factors per incident. Different things seemed to matter when fatal accidents were compared with non-fatal accidents. For example, bad work habits were less likely to result in fatal accidents, but organisational issues and co-worker problems were more common. They also looked at where the injuries occurred on the body. Fatal injuries were more often to the head and trunk, whereas non-fatal ones were usually to the upper arm. As for who was supervising or not supervising at the time of the accident, that mattered, too. Fatal accidents often happen when there is no supervisor around. Victims of fatal accidents were usually working more independently. Interestingly, almost half the fatal accidents occurred in places where, according to colleagues and supervisors, accidents had happened before.

As it turned out, the study showed that fatal and non-fatal accidents were not just the same things with different outcomes. They were different altogether. The findings back up Petersen's idea that different factors cause fatal and non-fatal accidents, debunking the old theory that they're just variations of the same thing, per the Heinrich model.

Another interesting take on statistics or metrics potentially being skewed comes from an article published by the Flight Safety Foundation (2020) on passenger mortality risk estimates, providing perspectives about airline safety. It is commonly known that everyone pays much attention to aeroplane safety. For example, the big news story of the TWA Flight 800 crash in

1996 in the United States even outshone AIDS, car accidents, and cancer in media coverage. People also change their flying habits when they hear about these tragedies. After one crash, ticket sales for a particular type of aeroplane plummeted by 36%.

There are many ways people try to measure how dangerous flying is. Some say, 'There were only 0.2 fatal accidents per 100,000 flying hours from 1993 to 1996,' which sounds good, but that statistic doesn't tell us if 1 or 300 people died in those accidents. Nor does it consider that most accidents happen during take-off and landing, not while cruising at 30,000 feet. Then there are the 'report cards' that rate airlines' safety. A score above 90 gets you an A, and below 60 is an F. Sounds simple, but there's a hitch: all fatal accidents weigh the same in these scores, whether one person dies or everyone onboard does. Also, if you're a smaller airline, one bad day could plummet your grade. There is also discussion around 'deaths per enplanement,' which refers to how many passengers died compared to how many flew. The issue with this is that a crash that kills everyone on board isn't three times worse if there are 150 passengers instead of 50. What if an airline's emergency procedures were reasonable and saved most people in a crash? That wouldn't show up in this kind of statistic.

The Q-statistic also needs to be factored in. A Q-statistic is a number that asks, 'If you pick a flight at random, what's the chance you won't make it?' It's a straightforward way to look at risk and even consider how many passengers die in each accident. But it's not perfect. Because fatal airline accidents are rare, the Q-statistic can swing wildly if just one crash happens, so figuring out whether an airline is safe is not straightforward. Some studies, like one by the FAA (1996), look at all the mishaps, not just the deadly ones. They believe bad luck can turn a minor incident into a fatal accident.

It could be assumed that you get a broader view by looking at more data. But that line of thinking is flawed. First, not every country has detailed records of minor incidents, and saying something is all down to luck is over-simplifying things. The curveball here is that the stats show that airlines with more minor incidents don't necessarily have more fatal accidents. Minor incidents doubled in the '70s and '80s, but the death rate didn't. It went down. When data on fatal and non-fatal incidents was compared, there was no apparent link to suggest one leads to the other. This also demonstrates that using minor incidents as a 'warning sign' for fatal accidents is unreliable and does not support Heinrich's pyramid.

Another line of thinking is that specific airlines might have more accidents than others based on certain factors. Or more flights might mean more chances for things to go wrong. Unfortunately, this is not the case. The study showed that airlines with more incidents sometimes turn out to be safer in the long run. It could all be a game of chance, as if you were flipping a coin. You might get heads ten times in a row, but that doesn't mean the next one will be heads, too. Some studies showed no significant safety difference between major US airlines over ten years. For example, one airline might have had a bad year, but that could even out over time. Again, as if you're flipping coins, sometimes you get a string of heads or tails, but it balances out eventually. In short, measuring airline safety isn't a one-size-fits-all thing. It's complex, and what happened in the past might not be a perfect predictor of the future.

In the United States, there are three kinds of airlines: big national ones, regional ones that stick to certain areas, and newer ones that started flying after industry rules changed. Think of regional companies as being like Alaska Airlines or Southwest and newcomers like Air South and Tower Air. Various research studies have tried to determine if one type of airline is safer than the other; however, what has been learned is that it's not as clear-cut as we'd like to think. The research looked at fatalities over the years and found the numbers too small to make any grand claims. Also, the risk of something tragic happening on a flight hasn't gone up or down over the years. The data can be misleading. For example, almost a quarter of all major crashes between 1977 and 1996 happened in 1987. The whole 'safety trend' could flip if you slice the data differently.

Smaller, non-jet commuter flights were also factored in. They had a higher rate of fatal incidents than the big jet flights, although fewer overall. It was unclear whether they were genuinely riskier or dealing with different conditions like small rural airports. Even airlines from developing countries tend to have higher risks. But, like everything else in this safety puzzle, the reasons could be varied and complicated. Flights from advanced countries like the US and Japan have a lower risk of accidents. But the safety stats are pretty similar when you look at flights from advanced and developing countries – like a flight from Paris to Karachi or Miami to Caracas.

For a bit more context, the data from 1987 to 1996 showed that the chance of something tragic happening was about 1 in 600,000 for these international flights, regardless of where the airline was based. In simpler terms, it doesn't matter if you're flying on an airline from an advanced country or a developing one on these specific routes. But, when flights within developing countries or between advanced countries were considered, those were a different story. Flights within developing countries had a higher risk, and those within advanced countries had a lower risk. However, when airlines worldwide shared the same routes, the safety numbers did not favour one. What this means is that the safety of your flight might have more to do with where you're flying to than which airline you're on, and if you were to fly every day, you'd have to wait around 19,000 years on average before facing a fatal accident, according to the math.

Another industry that has been struggling with quantifying safety for years is the construction industry. In Great Britain, for example, construction accounts for one-third of all work-related fatalities. That does not even consider illnesses or injuries. The number of deaths has slightly declined over the years, but the rate of significant injuries has recently increased. The most common mechanisms include falling from heights, inadequate housekeeping resulting in falls, and getting hit by moving or falling objects. The point to note here is that our numbers might be the tip of the iceberg. Many injuries aren't even reported, meaning that what we see is just a fraction of the accurate picture. Research shows that safety issues don't happen in a vacuum. It's not just one mistake or person – it's a web of factors. Poor planning, workplace errors, and even how complex the work is can all add to the safety risk. There is some data on this, but there's still much to learn.

Haslam et al. (2005) published research that looked at:

- The immediate events that have led to some nasty non-fatal accidents.
- What circumstances allowed these accidents to happen in the first place?
- How do these findings align with previous research?
- Most importantly, what lessons can we learn to make the construction industry safer?

The authors started with some focus groups, chatting with several people from the industry to figure out what safety problems need to be looked into. They then did a deep dive into about 100 real-life accidents that had happened recently. The authors got access to these accidents from companies that were willing to help. The goal was to cover various construction types and accident categories. For example, a high-rise building project accident might happen for different reasons than one in a residential home. The only rule was that they couldn't examine accidents already being checked out by health and safety authorities. Instead, they focused on less severe incidents and only considered accidents in the previous two months in which everyone involved was still around and willing to talk. The authors talked to the people in the accident and their managers to collect the data. They examined the accident location and reviewed documents that could give more insight. Experts in construction and human factors on the team then reviewed all this information and suggested the next steps for more research.

The reviewed data taught quite a lot about the various projects and how well they ran. For example, some projects were on time, whereas others were delayed. They even discovered that many of these less severe accidents could have been much worse under slightly different circumstances. The study also highlighted what was most often involved in these accidents: tools, machinery, or even building materials. Sometimes, the accidents were due to the way the construction site was set up. For example, half the accidents happened because the site standards were unsatisfactory. Factors like limited space and bad site planning were a problem. In one incident, a delivery vehicle tipped because there wasn't enough room to stabilise it correctly. Nearly all the accidents had an issue with risk management.

Some organisations in the construction industry are focused on designing safer sites, but most aren't there yet. Sometimes, it's about the budget. Other times, it's about not knowing better. There are ways to make things safer through better design, but the industry is slow to adopt these changes. Project management also had its own set of issues. Sometimes, it was unclear who was responsible for what, especially when multiple contractors were involved. Then, when poor planning was added into the mix, workers rushed, increasing the risk of an injury.

Interestingly, the maritime industry also has similar challenges. Regulations are the rule book that helps companies align with society's values. Think of them as a counterbalance to the intense competition that could otherwise lead to, let's say, cutting corners on safety or exploiting workers. This is super important in the maritime industry, which has been around forever and has its fair share of risks. The international safety management (ISM) code, designed by the International Maritime Organisation, supports the legislation. This code lets ship owners develop their safety management systems (SMS). This is important because, in a way, safety regulations in maritime transportation are joined at the hip to these SMS. The code is aimed at upping safety and environmental responsibility. Norway got on board in 1995, and it's been part of its Ship Safety Act since (Størkersen et al., 2017).

Norway has a long coastline and a deep-rooted history at sea. It has 402 companies running everything from ferries to cruise ships, and they're a significant part of the economy, responsible for around €1.442 billion in revenue and employing some 10,000 people. Between 2000 and 2014, 938 ship accidents and 2,704 injuries were reported. While fewer personal injuries have occurred over the years, the number of ship accidents has increased. An exciting study by Størkersen et al. (2017) dove into this topic. The study tried to determine why this was happening and whether the ISM code had anything to do with it.

Størkersen, Antonsen, and Kongsvik interviewed people involved in Norwegian maritime passenger transport. As it turns out, the ISM code is good for reducing individual accidents, but other factors in the industry lessen its impact when it comes to ship accidents. The research peeled back the layers to show how the ISM code is implemented, from top-level regulations to what's happening on the ships. The goal was to determine how these regulations affected safety efforts, both positively and negatively.

The Norwegian counties outsource maritime routes to private companies. The law demands they pick the cheapest option if all other things are equal. Companies are then highly motivated to cut costs, sometimes at the expense of safety measures. If a ship's late, the company even gets fined. These rules sometimes set the stage for safety to take a back seat (Størkersen et al., 2017).

The research showed that the ISM code has made maritime activities safer and more environmentally friendly. But there were some hiccups. The push for self-regulation led to a ton of paperwork and complicated procedures. This sometimes confused employees and clashed with their hands-on experience and common sense (Størkersen et al., 2017).

When the researchers interviewed 47 people involved in maritime operations, including the Norwegian Maritime Authority, the Norwegian Coastal Administration, and various high-speed boat companies, they determined that the ISM was generally considered good. It was believed to have been responsible for companies stepping up their safety game. However, some were a bit worried that the paperwork was too much. They believed this could distract crews from their primary job: keeping the ship safe. There was also concern that some safety efforts might have been 'window dressing' – looking good on paper but not making a real impact (Størkersen et al., 2017). Management at the boat companies saw the ISM code as formalising their safety work. They recognised that safety comes at a cost, and the code provides a basis to justify these expenses. However, they also felt tension between the old way of doing things and the new, paperwork-heavy approach (Størkersen et al., 2017). The people actually on the boats had mixed feelings. Some thought the ISM code hadn't changed much; others thought it was much bureaucracy with little real-world benefit. The crew members believed that the safety management systems (SMS) were too complicated and hard to follow, sometimes leading them to go with what they knew (Størkersen et al., 2017).

When you combine all these insights, you start to see a fuller picture. Some aspects of the ISM code help mitigate the risk of injuries (think slips, falls, etc.), but it's more complicated when mitigating the more considerable risks that can lead to ship accidents. On one hand, these rules have contributed to better safety habits, which could help reduce the number of injuries. On the other hand, they've created a situation in which paperwork and compliance get in the way of running the ship safely. It's interesting to note how one set of regulations can have different impacts. According to the study, this might be because there's a disconnect between what the regulations intend to do and how they're implemented in the maritime industry (Størkersen et al., 2017).

Norway's maritime industry shows that safety regulations like the ISM code deliver mixed results. They get people thinking about safety more, but they're also creating a lot of paperwork and stress, which could distract from the main goal – keeping people and ships safe.

5

BUT I AM ONLY INJURING MY SUB-CONTRACTORS!

The safety of subcontractor employees in high-risk workplaces has been studied for a while, dating back to the '80s and even more recently (Lamare et al., 2015; Quinlan, 2014). Subcontractor workers are more likely to get into accidents than regular employees (Muzaffar et al., 2013). Most previous research has examined how the principal or prime contractor chooses and manages these subcontractors. But even when they have systems to manage subcontractors, accidents still happen more often among them.

Subcontractor employees often find themselves in riskier situations. They might not always report injuries, making it seem like everything is fine on the surface (Kenny & Bezuidenhout, 1999). But when you dig a bit deeper, you'll see that they experience more severe injuries and fatalities than the regular crew. Think of it this way: Imagine you have two groups of friends playing different sports. Group A plays football, and Group B plays rugby. Group A rarely talks about getting hurt, but when they do, it's usually minor stuff like a scraped knee. Group B, the rugby players, doesn't talk much about their injuries either, but when they do, it's often something more

DOI: 10.4324/9781003177845-5

serious, like a sprained ankle or worse. Subcontractor employees are a bit like those rugby players. They face a greater risk of work-related diseases and tend to be absent from work more often (Min et al., 2013). They also have a higher proportion of fatal injuries (Muzaffar et al., 2013).

A real-life example is the Pike River Coal Mine disaster in 2005. It was a catastrophic event with multiple methane gas explosions, leading to the tragic loss of 29 lives. Heartbreakingly, 13 who perished were subcontractor employees (Lamare et al., 2015). When investigators looked into what went wrong, they found some glaring issues. The site management didn't keep proper records of where the contractors were on-site. They even hired contractors who had no previous experience working in a mine. Perhaps the most shocking part was the lack of a solid safety management system for these contractor employees (Macfie, 2015).

The use of subcontracting has been on the rise for quite some time. In Australia alone, there were over one million independent contractors as of 2015 (Australian Bureau of Statistics, 2016). In the construction industry down under, subcontracting accounted for a massive $93.6 billion during the 2011–2012 fiscal year, making up 40.1% of the industry's total income (Australian Bureau of Statistics, 2013). Industries that rely heavily on subcontractors also tend to have some of the highest numbers of injuries and fatalities. Take mining in Australia, for example. Between 2007 and 2012, 36 mine workers died while on the job (Safe Work Australia, 2014). The construction industry isn't far behind, with 49 fatal incidents recorded between 2012 and 2013 (Safe Work Australia, 2014). Whilst there has been a slight decline in these numbers, they remain well above the national average. For instance, the mining industry had an incident rate of 3.84 per 100,000 workers, which is 70% higher than the national rate for all industries (Safe Work Australia, 2014).

In the United States construction industry, one study found that an increase in subcontracting was directly linked to a rise in injury rates on-site (Azari-Rad, 2015). Another investigation noted that a staggering 59% of deaths caused by trench collapses involved subcontractors (Suruda et al., 2002). In Turkey, they also studied shipyard-related fatalities and found that 25% of fatal workplace incidents were related to subcontractors (Barlas, 2012).

To put things into perspective, imagine you're at a job fair, and two companies offer you work. One company is super competitive and wants the best people at the lowest cost. The other one offers a steady job with no worries about income. Which one would you choose? For subcontractor employees, it's a bit like this job fair. They face two types of pressure: economic pressure from the contracting company's perspective and job and income insecurity from their own viewpoint. The hiring process can be intense. Sometimes, even before the bidding process starts, clients and principal contractors assess to determine who should be considered. The lowest bidder usually gets the contract. The catch is that pricing often takes the top spot on the checklist when awarding contracts. This intense competition and the race to lower costs can make subcontractors prioritise safety lower on their list (Lamare et al., 2015; Min et al., 2013).

Researchers suggest that considering subcontractors' safety track records when choosing them should be a bigger deal (Huang & Hinze, 2006; Ng et al., 2005; Roughton, 1995). Some companies use safety metrics for contractor qualification and bidding, but whether this genuinely rewards safety is unclear. There have been only a few cases in which safety improved after implementing this strategy. The other side note is that subcontractors often face delays in getting paid. In a survey, a whopping 89% of subcontractors said their payments were delayed by more than 45 days (Arditi & Chotibhongs, 2005). Some contracts even include clauses such as 'pay-when-paid' or 'pay-if-paid' to postpone subcontractor payments (Kirksey, 1992; Uher, 1991).

As a contractor, reporting an injury doesn't always bode well for job security (Kenny & Bezuidenhout, 1999; Quinlan & Bohle, 2004). There are instances when injuries go unreported or are downplayed. For example, on North Sea oil platforms, subcontractors often hesitate to report incidents and injuries because they fear these reports might affect their chances of getting more work (Collinson, 1999). In South African mines, some injuries go unreported, and contract employees continue to work with injuries (Crush et al., 2001). Subcontractor employees also tend to work longer hours and have multiple jobs to maintain financial stability, which can increase their safety risks (Lamare et al., 2015).

Subcontractors are less protected by safety management systems (SMS)

Management has a prominent role to play in helping subcontractors keep safe. Their commitment to safety is massive, especially in high-risk industries (Ali et al., 2009; Butz & Leslie, 2001; Sawacha et al., 1999; Zohar,

1980; Zohar & Luria, 2003). Usually, in high-risk industries such as mining or construction, there are safety systems, regular training, and the right gear to protect workers. Regular employees trust the management team regarding safety (Sætren & Laumann, 2015). But things are a little different for sub-contractor employees. They often face confusion about who's responsible for their safety, especially regarding which safety management system to follow. Smaller subcontracting companies might not have safety systems and rely on the main contractor's system. The problem is that managing multiple systems across different locations can be a headache (Bahn, 2012). Subcontractors will usually follow the principal's system. This can lead to issues, especially if the contractor's safety team lacks experience (Huang & Hinze, 2006), or there is poor communication between different contractor teams working on the same site (Loosemore & Andonakis, 2007; Simon & Piquard, 1991). It's like trying to play a symphony with musicians who can't read the same sheet music. Another concern is the lack of proper training the site owner or principal provides. Not having the proper training can hurt subcontractors' safety records (Hinze & Gambatese, 2003; Rebitzer, 1995). Subcontractors in construction and mining are often small to medium-size companies, some with fewer than 20 employees. Many cite high costs, language barriers, and resistance to change for not implementing proper occupational health and safety (OHS) standards and education (Loosemore & Andonakis, 2007).

Enforcing OHS regulations can be challenging when it comes to subcontractors, especially labour-hire subcontractors (Johnstone & Quinlan, 2006). Government regulations often treat regular employees and contractor employees the same, but the nature of temporary employment makes it challenging to ensure compliance. It's like trying to ensure everyone follows the same rules at a massive music festival – not an easy task. Union membership also plays a role in protection. Regular union employees have more safety nets than contractor employees who lack union access (Gillen et al., 2002). This might be because more subcontractors hire casual employees (Rousseau & Libuser, 1997). Sometimes, subcontractors do their best to comply with OHS regulations, but not all succeed. It could be due to the pressure to finish a project on time, a lack of awareness about specific regulations (Hislop, 1999), or a willingness to undertake riskier work.

In the African gold mines, subcontractor employees, often hired temporarily, are typically assigned to work in the most perilous parts of the mines.

These are areas where regular miners, backed by their unions, have said they would not work. To make matters more challenging, these temporary employees often work longer hours because mine management doesn't keep as close an eye on their shifts as they do for full employees (Crush et al., 2001).

In Sweden, they've noticed something interesting in the mining sector. There's a lack of data on subcontractor workers' safety incidents, but what they do have suggests that these workers have a higher rate of injuries which are more severe. Plus, they seem to be doing different tasks under different conditions than the permanent employees when accidents happen (Blank et al., 1995).

The nuclear industry is also known for its high-risk nature. Even here, subcontractor employees bear a more substantial share of risk. In Japanese nuclear facilities, subcontractor employees often handle the riskiest tasks in the most dangerous sections of the plant and are exposed to higher radiation levels (Laqua et al., 1997). An eye-opening example comes from the Fukushima nuclear plant, where subcontractor employees doing the cleanup work were exposed to radiation levels more than eighteen times higher than regular employees (Jobin, 2011). Specialist subcontractors are frequently tasked with the most hazardous phases of nuclear decommissioning – these tasks are critical but also the most time consuming (Owen et al., 2013). A similar pattern emerges in the French nuclear industry, where subcontractor employees work in higher-risk areas with tight schedules and greater demands (Thebaud-Mony et al., 2011).

Contingent workers in the US share some unique insights. Subcontractors often don't feel like part of the team. They sometimes struggle to get a day off; face payment issues; and, as previously noted, are assigned more challenging, riskier jobs (Padavic, 2005). In France, a study by Roquelaure et al. (2012) found that temporary workers didn't report significantly more injuries than permanent employees. However, they faced more time constraints and repetitive work, which increased their perceived risk of musculoskeletal injuries. A study based on Korean working conditions data revealed that subcontracted employees often worked in more hazardous conditions. This led to higher rates of absenteeism and an increased risk of occupational diseases (Min et al., 2013).

It would be easy to say, 'But it's just our contractors,' and focus the blame entirely on those called in to assist organisations with a contract; however, case studies provided in an article by Valluru et al. (2020) demonstrate the challenges that contractors face in their daily work and the impact this has on reporting and recording of injuries.

Case study 1 – the tyre burst tragedy

In this case, a truck driver was in a fatal accident when one of his tyres burst. The driver worked for a mining services firm that the mine operator subcontracted. Even though the mining services firm was relatively small, it had grown over time and operated under the mine operator's safety management system.

With 21 months of experience, the driver was responsible for hauling coal and performing minor truck maintenance tasks. On the day of the incident, he took over the truck from a night shift driver, who believed it was safe. However, the incoming driver noticed low tyre pressure and wanted to change it. His supervisor, though experienced in driving, had limited experience changing tyres but approved the task.

Changing mine truck tyres is hazardous due to the high pressures involved, and an audit had updated the standard operating procedures (SOP). Unfortunately, the mine operator didn't communicate these changes to the mining services firm. The truck driver started working on the wheel without proper training or understanding of the updated procedures, leading to a catastrophic explosion.

Case study 2 – the loader operator's nightmare

In this tragic incident, a loader operator was fatally crushed by the cab of the loader he was driving. The subcontractor operator had been hired for bulk sampling operations on a mine site. He was unfamiliar with the loader supplied and had no previous experience with it. The loader had critical mechanical issues, including a malfunctioning horn and ineffective brakes.

When the loader's engine stopped, the operator attempted to halt the vehicle by steering it into a bank, but it tipped over and crushed him. This incident highlighted issues like inadequate equipment maintenance and the absence of operator competency records. Moreover, the site's safety management plan didn't include plant operators, exposing a gap in safety procedures.

Case study 3 – the hopper repair gone wrong

This case involves a subcontractor employee who was fatally crushed while repairing a quarry hopper. The employee was an apprentice boilermaker, and he and a colleague attempted the hopper repair without proper induction or training. They were unaware of standing instructions to remove hopper doors before maintenance.

The lack of an operating manual, limited tools, and a site safety executive who didn't supervise the work contributed to this tragic accident. It highlights the importance of proper training, supervision, and clear safety instructions for subcontractor employees.

Case study 4 – the water truck tragedy

In this case, a subcontractor truck operator died when his truck crashed through a bund wall during a watering activity. The operator had extensive experience but faced challenges due to inadequate brakes, poor truck maintenance, and a lack of formal contractor management plans.

This incident emphasised the need for regular equipment maintenance, effective safety plans, and communication between principal contractors and subcontractors.

Case study 5 – the pumping mishap

This case involves a subcontractor employee who was injured fatally when a pump fell on him. Two subcontractor teams were assigned to complete interconnected tasks, but miscommunication and failure to follow safety protocols led to the accident. It highlights the importance of clear communication, adherence to safety procedures, and proper equipment use.

Case study 6 – the underground mine tragedy

In this case, an underground mine worker suffered a fatal injury when struck by a rib spall. The incident resulted from several factors, including missing rib bolts, improper inspections, and inadequate communication between subcontractor employees and mine operators. This case underscores the need for regular safety inspections, effective communication, and proper support for subcontractor employees.

These case studies highlight the fact that, whilst principal contractors' systems and processes may seem acceptable, subcontractors often face challenges that lead to various employee risks. This lack of management leads to disaster when not managed or controlled appropriately, with assistance from all parties.

6

MEASURING NOTHING IN 'ULTRA-SAFE' INDUSTRIES

In high-reliability industries, safety is not just a priority but a fundamental necessity. Lofquist (2010) completed a transformative study focusing on the ultra-safe civil aviation industry during significant organisational change. The research was based on a three-year case study of Avinor, the Norwegian air navigation services provider. The study aimed to determine how deliberate large-scale or strategic changes impact safety in high-reliability organisations (HROs). The context for this research was the government's corporatisation of air navigation services in Norway. This initiative aimed to maintain or improve safety while transitioning to a corporate structure without disrupting customer services. However, traditional safety metrics, such as incident and accident reporting, did not adequately capture the harmful effects of these changes, both during and after the change process.

The research paper had two main objectives. First, it addressed the challenge of measuring safety outcomes during organisational change in ultra-safe industries using traditional safety metrics. The issues observed

DOI: 10.4324/9781003177845-6

in the Avinor case study are argued to apply to the civil aviation industry and other high-risk sectors. Secondly, the paper proposed an expanded integrated safety model based on existing safety management systems (SMS) widely used in civil aviation. This model presents safety as an emergent property of a complex socio-technical system within an organisational culture over three temporal phases, going beyond traditional reactive measures and providing a more comprehensive view of evolving organisational safety. The study's background is rooted in concerns raised in 2005 about reduced safety margins in the Norwegian air transport sector due to strategic and incremental changes in the industry. A review by the Norwegian Accident Investigation Board (HSLB) suggested that while safety levels remained high, traditional accident statistics were insufficient to assess the true impact of these changes on flight safety. This raised questions about whether safety is merely an outcome or an emergent property of a complex system, especially during strategic change.

The paper also discussed the concept of corporatisation in the civil aviation industry. This process began in the late 1960s, aiming to privatise or semi-privatise national civil aviation activities while maintaining high safety standards. This transition places added responsibility on organisational leaders to balance financial performance with safety outcomes in an industry in which incidents and accidents are rare by design. Conflicting organisational goals during this process can impact safety outcomes, often not immediately evident but potentially catastrophic in the long term. The literature review highlighted the need for a systems approach to studying safety management systems in high-risk industries, moving beyond traditional reactive measures. It mentions the importance of interdisciplinary contributions from sociology, psychology, engineering, and other fields to understanding safety better. The study argues that to achieve a genuine systems perspective on safety, it is essential to involve system operators and business leaders responsible for safety outcomes.

In examining the complexity of civil aviation as a 'social-technical system,' it's essential to understand its unique challenges. While safety studies often focus on large machine-bureaucratic organisations like nuclear power and chemical industries, civil aviation has received comparatively less attention. However, the aviation industry has seen significant growth, technological advancements, and increased complexity in recent years. Civil aviation is best described as a complex system comprising overlapping socio-technical systems in a highly competitive business environment, with safety as a primary but not sole priority. Unlike tightly coupled machine-bureaucratic systems, civil aviation relies heavily on human interaction and is profoundly influenced by human variability. Air traffic control is notable for relying on controllers' skills, making human variation integral to system design and performance.

Loose coupling, akin to feedback with a delay in system dynamics, adds complexity to understanding causal relationships over time. These complexities are often overlooked in incident investigations focusing on the most apparent 'root causes.' This holds during dynamic periods of change when system outcomes diverge from the original design yet remain safe, thanks to operator adaptation, regulations, or built-in resilience. Complex causal relationships become even more challenging to identify when examining overlapping safety management systems, as external factors can obscure system deficiencies. The civil aviation industry exemplifies this complexity, with multiple high-reliability organisations offering overlapping safeguards that can both prevent and contribute to disasters. These 'safety nets' can lead to the development of latent conditions that hide potential disasters in unexpected ways.

Measuring safety in civil aviation is a significant challenge. Relying primarily on catastrophic outcomes as safety metrics is problematic because the likelihood of such events has decreased substantially, approaching ultra-safe levels. Emphasising disastrous events alone gives undue weight to their consequences and ignores unintended outcomes that could have had catastrophic results but did not. Safety is often defined and measured more by its absence than its presence, as the Norwegian Accident Investigation Board noted in its study of the Norwegian civil aviation industry. Safety in civil aviation operates in three phases: proactive, interactive, and reactive.

Proactive measures involve system design and redesign, considering operational assumptions and technological constraints. These assumptions evolve, creating gaps between expected and actual performance. The interactive phase deals with real-time system operation, in which internal and external factors and human variability are crucial. It's a critical phase for improving safety outcomes, particularly during deliberate changes. Organisational culture and leadership significantly influence the interactive phase. Safety processes in this phase depend on robust interfaces between system designers, operators, and managers, facilitating early identification of discrepancies and latent conditions. The reactive phase involves learning from undesired events and enabling system corrections through restrictive measures or system redesign. Restrictive measures alone may not yield significant safety improvements. A balanced approach to measuring safety in complex systems like civil aviation involves monitoring safety across these three phases, with leading and lagging indicators, quantitative and qualitative measures, and operator perceptions. Such an integrated safety measurement system provides a comprehensive view of an organisation's safety state and supports proactive decision-making in a changing environment.

In the Avinor study, a three-year longitudinal case study was conducted to examine a strategic change initiative called Take-Off 05. The research used a mixed-method approach, combining quantitative and qualitative data from various sources over two years. The primary aim was to investigate how this change initiative impacted safety perceptions in the safety management system model. Data collection occurred at two specific points: just before the implementation of Take-Off 05 and at the mid-point. Two levels of aggregation, including four embedded cases and the entire air traffic controller/ATC assistant population, were used to minimise potential bias effects. The first set of quantitative data came from a leadership questionnaire administered in Avinor's air traffic control centers (ATCCs) in December 2002, three months before the announcement of Take-Off 05. This questionnaire was repeated at the mid-point of the initiative in November/December 2004. Some ATCCs had merged during this time, while others faced closure or integration. Participation rates varied across locations. A semi-structured interview protocol was developed after analysing leadership questionnaire results and observations. These interviews focused on four critical latent concepts related to safety perceptions during organisational change: leadership commitment to safety, safety culture, attitudes toward organisational change, and individual safety perceptions. These interviews, conducted at each site (ten at each location), were transcribed and coded using NVivo 7 software, (NVivo is a qualitative data analysis (QDA) computer software package.) The study focused on the three remaining ATCCs, which were similar in size and function and covered 100% of Norway's en-route civil aviation structure.

The second quantitative dataset was obtained from the Norwegian Transportation Safety Board (HSLB) and included responses from various groups in the civil aviation industry. Only responses from air traffic controllers and ATC assistants were used to conduct a confirmatory factor analysis on a conceptual structural equation model. The findings indicated that individual reactions to the change process and perceptions of local leadership and the safety climate varied depending on the phase of the change each unit experienced. However, perceptions of top leadership commitment decreased significantly across all units. The study also presented a conceptual structural equation model based on four latent concepts related to change and safety attitudes. This model showed strong correlations between independent constructs and the primary dependent construct (perception of safety), with leadership commitment influencing safety perceptions indirectly through attitude toward change.

In summary, the research demonstrated a significant decrease in individual safety perceptions during organisational change, highlighting the challenge of measuring safety outcomes in ultra-safe industries during periods of change. The study emphasised the need for integrated safety management systems and proactive measures to detect potential system weaknesses during change processes. Further research should explore how organisations handle and implement change in high-risk industries, especially when safety incidents are infrequent.

7

DON'T TALK ABOUT SAFETY CULTURE – TALK ABOUT SAFETY CLIMATE!

According to the Bureau of Statistics, in the United States, an estimated 4.1 million work-related injuries and illnesses are reported annually, with a rate of about 4.4 cases for every 100 full-time-equivalent workers. However, recent research indicates that these numbers may not fully capture the extent of non-fatal occupational injuries. This underestimation can occur due to organisations failing to record employee injuries and illnesses or employees choosing not to report them – factors like organisational size, industry sector, management responsiveness, and safety climate influence organisational-level under-reporting. In contrast, individual-level under-reporting can be driven by fear of reprisals or the belief that injuries are a normal part of specific jobs.

Despite advances in understanding the factors influencing under-reporting, little research has explored the impact of organisational safety climate on the accuracy of employee reports of workplace injuries. Additionally,

DOI: 10.4324/9781003177845-7

there is limited research on how supervisory enforcement of safety practices affects reporting behaviours. A study by Probst and Estrada (2010) aimed to address these gaps by examining under-reporting among workers in above-average risk industries, investigating the role of safety climate in experienced accidents and under-reporting. The study drew data from a diverse pool of 425 employees representing various organisations. These organisations included a light manufacturing firm with 136 participants, a small heating and cooling company with 22 participants, seven dental clinics operating under a regional dental care organisation with 77 participants, a pulp and paper mill with 128 participants, and several restaurants in the hospitality industry with 62 participants. The selection of these industries and organisations was deliberate because they encompassed a broad spectrum of sectors, each exhibiting at least an average risk of employee injuries. For context, consider the following: In 2007, the dental industry reported approximately 2.4 injuries per 100 full-time employees, whereas the heating and cooling sector recorded around 6.9 injuries per 100 full-time employees, according to the US Bureau of Labour Statistics.

To ensure that employees were well informed about the study, they were briefed in advance through regular meetings and official invitation letters sent by the researchers. Additionally, the study received endorsements from upper management, and union officials at the mill, the only unionised site, supported them. During the data collection process, researchers distributed surveys containing relevant measures to groups of employees. These survey sessions were conducted during employees' regular working shifts. The surveys were accompanied by a cover letter outlining the study's general objectives and detailing the measures in place to safeguard the confidentiality of respondents. A comprehensive effort was made to reach all employees in these organisations by scheduling survey sessions during different work shifts.

Furthermore, employees were given time off during regular working hours to complete the surveys. While participation rates were estimated to be between 80% and 90%, exact figures remained elusive due to potential variations in the number of employees present on the days the surveys were administered. This variance was particularly noticeable in the restaurant settings and dental clinics.

Regarding demographics, most participants were male Caucasians, accounting for 65% of the sample, with an average tenure of 8.60 years (standard deviation = 9.47 years). A significant portion of employees (54%) had a high school education or less, while 33% had completed some college education. Respondents' median and most frequently occurring age category was 35–39. Analysing the demographics in each organisation revealed that the survey respondents closely mirrored the overall workforce composition. Furthermore, when comparing the gender and racial makeup of the industry samples to the US census data for broadly defined industry categories, the samples appeared to be reasonably representative in terms of these demographic characteristics. It's worth noting, however, that the restaurant sample had a higher proportion of males and Caucasians than the industry's general population.

What was measured?

- *Accidents*: The study employed a scale that Smecko and Hayes (1999) developed to assess workplace accidents. This scale comprised two items: one focused on experienced and reported accidents and the other on experienced but unreported accidents. For instance, think of an employee working for a construction company who had to report any safety-related incidents or accidents they encountered. Participants were required to indicate how many safety accidents they had experienced and subsequently reported to their supervisor, as well as how many accidents they had experienced but chosen not to report over the past 12 months. The range for reported accidents varied from 0 (the most common response) to a high of 13, while unreported accidents ranged from 0 to a high of 37.
- *Organisational safety climate*: The study assessed organisational safety climate using 16 items borrowed from Griffin and Neal's (2000) scale. These items were designed to measure four dimensions of safety climate: management values, safety communication, safety training, and safety systems. Participants rated these items on a Likert-type scale, ranging from strongly disagree (1) to strongly agree (7). The scores were averaged to create a composite score, in which higher scores indicated a more favourable safety climate. Consider, for example, a retail company where employees were asked to rate the importance of safety in their workplace and their level of agreement with safety communication and training statements.

The study's results demonstrated that a positive organisational safety climate was associated with fewer reported and unreported accidents. To illustrate, consider a hospital where a positive safety climate, emphasising workplace health and safety, was linked to decreased reported and unreported incidents among healthcare workers. Additionally, a positive safety climate was related to a reduced number of total experienced accidents. The study also delved into the reasons behind under-reporting accidents and the consequences of reporting accidents. Employees often chose not to report accidents because they believed they could resolve the issue independently, wanted to avoid follow-up interviews, or held the perception that nothing would be done to rectify the problem. Furthermore, some respondents expressed concerns about the impact on their safety scorecard or the company's 'accident-free record.' For instance, workers might avoid reporting minor incidents at a construction company to avoid time-consuming interviews or because they believed management wouldn't take action.

This study investigated employee under-reporting of accidents while exploring whether organisational safety climate and supervisor enforcement of safety policies play a role in such under-reporting. The results, drawn from samples across five different industries, consistently demonstrated that under-reporting accidents is a prevalent issue. Prior research had suggested that up to 68% of workplace accidents and injuries were not documented in national injury surveillance systems like OSHA and BLS (Rosenman et al., 2006). This study found that 71% of experienced accidents went unreported, aligning with the earlier findings. Furthermore, this study revealed that employees' perceptions of organisational safety climate could predict the extent of under-reporting. When employees viewed their organisational safety climate positively, the tendency to under-report decreased significantly (with a ratio of unreported to reported accidents at 1.46:1). Conversely, in organisations where employees perceived a poor safety climate, the ratio of unreported to reported accidents exceeded 3:1. This emphasises the importance of evaluating reported and unreported accidents independently, as they may have distinct influencing factors.

To better understand these results, supplementary data was collected from mill employees, shedding light on the reasons for not reporting accidents and their perceived consequences. These findings corresponded with the safety climate analysis, suggesting that explanations for not reporting accidents related to significant dimensions of the safety climate, such as management's value of safety, safety communication, and safety systems. It's noteworthy that nearly two-thirds of employees who reported accidents also reported experiencing negative consequences. Organisations often assert that the health and safety of their employees are top priorities. However, accidents and their under-reporting continue to be common problems. While previous research has shown that a positive organisational safety climate can reduce accidents, this study is the first to reveal its substantial connection to accident under-reporting.

Hence, organisations must recognise the benefits of fostering a positive safety climate and an environment in which accurate reporting is encouraged. Organisations might believe fewer reported accidents are beneficial in the short term (e.g. lower workers' compensation loss rates). Still, this study suggests that such practices may have detrimental long-term effects on employee health and safety. Addressing the contributory factors of injuries or accidents, rather than simply reducing their visibility, is vital. A positive safety climate relates to fewer workplace accidents and encourages accurate reporting when accidents occur. This is analogous to 'noisy fixing,' as discussed in Ancona and Bresman's book on building successful teams at work. At Toyota, for instance, if a car comes off the assembly line with a defective part, the issue is not quietly fixed but thoroughly examined to prevent its recurrence. Similarly, fostering a positive safety climate ensures that safety concerns and incidents are brought to the attention of coworkers and supervisors, addressing the contributory factors and mitigating their risks.

Finally, once the relationship between safety climate and under-reporting is firmly established, intervention studies can be designed to improve safety climate within organisations, subsequently assessing the extent to which this change reduces the under-reporting of accidents.

8

WHAT IF NOT EVERY INJURY IS PREVENTABLE?

In 2017, Fred Sherratt and Andrew Dainty in the UK asked themselves whether 'zero harm' influences managerial, administrative, or operative safety practices and, in turn, whether it affects safety outcomes (Sherratt & Dainty, 2017). It was high time. In the mid-2010s, many companies started changing their advertising of safety positions, calling them 'zero harm advisors,' so keenly felt was the virtuous obligation to signal a commitment to an injury-free workplace. The only number that mattered anymore was 0. However low an LTI rate was, it still wasn't good if it wasn't zero. Of course, you can unequivocally argue that it is virtuous to have as a goal not to hurt anyone. It is also a contractual employer obligation in many jurisdictions.

The suspicion, however, had long been that a commitment to zero might have unintended consequences or adverse effects which run directly counter to the objective it espouses. This is particularly the case if the commitment becomes a target and certainly when the achievement of that target is incentivised with bonuses – independent of where in the organisation they are awarded (Maslen & Hopkins, 2014):

DOI: 10.4324/9781003177845-8

In her research on Zero . . . Sherratt found that corporate programmes often position Zero as a tangible goal, a firm 'future perfect' reality, which can be counted and measured through many targets. Yet this utopia was challenged and even derided by the construction workers themselves, for whom the lived realities of their working lives tell them Zero is, and is likely to remain, a utopian fantasy, totally incompatible with the current challenges of production that they face daily. Worse still, it might stymie the open dialogue and learning culture widely acknowledged as a hallmark of a progressive safety discourse.

(Sherratt & Dainty, 2017, p. 3)

'Outlawing' incidents and injuries can inspire cynicism, does very little to improve safety (Donaldson, 2013; Sherratt, 2014), and could conceivably make things worse by driving under-reporting and risk secrecy (Dekker & Pitzer, 2016; Zadow et al., 2017). This is backed up by earlier studies into the relationship between non-fatal and fatal injuries incurred in workplace incidents (Mendelhoff & Burns, 2013; Salminen et al., 1992; Saloniemi & Oksanen, 1998).

To empirically test this in the 21st century, Sherratt and Dainty were able to research whether large construction organisations operating in the UK that had adopted Zero had, indeed, seen improvements over those that had not. They also had the opportunity to investigate whether they developed greater safety in practice and showed any evidence of beginning a process of 'innovating to Zero.' The researchers obtained fatal and major accident data for the UK construction industry from 2011–2012 to 2014–2015 under a Freedom of Information Request to the UK Health and Safety Executive (HSE). The significant accidents included injuries such as amputations, reductions in sight, crush injuries, significant burns, and loss of consciousness – all of which could be life changing. The FOI data also included the identity of the reporting company (which, in the case of accidents occurring on large construction sites, would most likely be the principal contractor).

Sherratt and Dainty analysed the safety programs of the top twenty companies, which accounted for about a quarter of all construction work in the UK. Nine of them had an explicit Zero policy in place. Six of these companies had a safety program that referenced Zero. The three others contained clear announcements around Zero as a target or specifically aimed to be 'incident- and injury-free.' Over the four years for which data was gathered and analysed, they found that:

- There were four fatal accidents at companies with Zero safety.
- There were zero fatal accidents at companies without Zero safety.

Not only that. Whether a company had explicitly committed to Zero also correlated with the number of significant injuries:

- There were 214 major injuries at companies with Zero safety.
- There were 135 major injuries at companies without Zero safety.

Of course, these figures were not corrected for variations in the volume of work carried out by the top twenty companies in the four years of the study. Without access to work hours, revenue was used as a proxy for volume and showed that:

- There were seven fatal or major accidents per billion turnovers for those with Zero.
- There were six major accidents per billion turnovers for those without Zero.

The research suggests the possibility of a 'Zero paradox.' More accidents and fatalities were occurring in companies working under a Zero banner. As a construction worker, you were marginally more likely to have a significant accident or incur a fatal injury while working on a large construction site operated by a contractor with any form of Zero. You were less likely to suffer such a significant accident or to be killed if you worked on a site without Zero. For construction on large UK sites, Zero meant a greater risk of injury (or death) in practice for the four years studied. As the researchers concluded:

> Whilst it must be acknowledged here that this data is limited to only 4 years, it nevertheless suggests that there is at least the potential for an increase in accidents following the introduction of Zero safety on sites. There is no guarantee that the implementation of Zero safety can ensure continued or even any reduction in accident rates overall, and indeed, it seems to have limited

impact in catalysing any significant step-change improvement in safety performance. . . . That Zero has not brought about effective change over the four years examined here suggests the realisation of some of the more critical challenges made against it. As Long argued, Zero in practice can easily foster the development of non-learning cultures, closing down debates around safety and a resultant focus on failure. Unfortunately, within this dataset, failure is all too apparent, and there has been a lack of coherent, evidenced improvements in safety performance for those operating under the banner of Zero.

(pp. 6–7)

Declaring a Zero vision, in other words, can reduce operational knowledge, lead to the manipulation of incident and injury figures, and restrict organisational learning. As we saw earlier in the book, when a measure becomes a target (in this case, zero), it stops being a measure. It just becomes a target that must be achieved at all costs. The decline in disclosure, honesty, and learning that happens as a result can increase the probability of significant accidents and fatalities. This is consistent with high-reliability organisational orthodoxy: creating climates of psychological safety in which getting bad news to the boss is encouraged is a crucial way to learn about and manage operational risk (Rochlin, 1993; Weick & Sutcliffe, 2007).

Of course, this is an argument that cannot be sustained by argument alone because what happens in the organisation committing to zero harm depends heavily on its cultural proclivities. There are historical examples of organisational cultures that had no tolerance whatsoever for any deviation, failure, incident, or even indication of any possible harm that actively incentivised assertive, immediate conversations about such indications and collective responsibility to create meaningful interventions or corrections. This was the case for Japan's 'zero-accident total participation' campaign in the 1960s, which predates the West's Zero Accident Vision (JICOSH, 1964). Pointing and calling out hazards on dedicated walks (in which workers were instructed exactly how to 'strike a pose with spirit, straighten themselves, and then briskly point to the hazard'), with workers taking responsibility for the collective good, worked in that cultural setting. There is, however, no data from that time about whether this helped prevent major incidents or fatalities.

The Zero paradox doesn't exist in the construction industry or manufacturing. Committing zero incidents/injuries in oil and gas doesn't prevent fatalities, either, the available data shows. In his comments on a 1998 gas explosion at an Esso plant in Victoria, which killed two people and injured eight, Hopkins (2001) wrote:

> Ironically, Esso's safety performance at the time, as measured by its Lost Time Injury Frequency Rate, was enviable. The previous year, 1997, had passed without a single lost time injury, and Esso Australia had won an industry award for this performance. It had completed five million work hours without a lost time injury to either an employee or contractor. LTI data are thus a measure of how well a company manages the minor hazards that result in routine injuries; they tell us nothing about how well significant hazards are managed. Moreover, firms typically attend to what is being measured at the expense of what is not. Thus, focusing on LTIs can make companies complacent about managing major hazards. This seems to have happened at Esso.
>
> (p. 4)

Other petrochemical accidents have elicited the same reflections. For example, the Chemical Safety Board found that the 'BP Texas City explosions were an example of a low-frequency, high-consequence catastrophic accident. Total recordable and lost time incident rates do not effectively predict a facility's risk for a catastrophic event' (CSB, 2007, p. 202). Another such case seems to be the 2010 Macondo (or Deepwater Horizon) well blowout in the Gulf of Mexico, which killed 11 people and caused the biggest oil spill in the history of humanity. It was preceded by a celebrated six years of injury-free and incident-free performance on the platform (or boat, really) (BP, 2010; Graham et al., 2011). A reported zero injuries or incidents preceded 11 deaths. In year-on-year data published by BP over the decade from 2005 through 2015, there were 82 fatalities in total (BP, 2017). Per year, these fatalities do correlate modestly with recorded incidents ($r1/40.59$), though this is not statistically significant ($p<.051$). And it failed to predict Macondo. As we saw in earlier chapters, achieving a zero target on the small stuff predicted nothing about the big stuff.

The idea of preventability

As discussed in Chapter 1, the European Enlightenment, a series of social, intellectual, and political movements during the late 17th and 18th centuries, emphasised human reason and a willingness to question authority and promoted a somewhat 'utopian' vision. The promise, or at least the idea, was that we were not condemned to live in a Hobbesian nightmare; society could be perfect. Suffering could perhaps be made redundant. In 1782, French Enlightenment thinker Marquis de Condorcet celebrated the idea he called 'so sweet.' People could be improved and society perfected with the help of 'those sciences, the object of which is man himself' (Hacking, 1990, p. 38). Measurement, mapping, and quantification were vital. A truly 'perfect' society was one in which interventions by the state and other authorities could bring harm and suffering down to zero.

Both the utopian aspiration and the focus on measurement of the Enlightenment show up in zero harm. This is particularly the case where zero harm has become a target – or, to be more nuanced, where the broad vision is reduced to a single lagging indicator such as LTI figures posted next to site entrances or advertised in annual reports. When placed in a business environment driven by bureaucratic organisation and competitive profit, the originally noble commitment can precisely undergo the sort of transformation Merton (1938) warned against, where it inevitably generates secondary consequences that counter its objectives. Confusion and cynicism among the workforce can be the result. As Sherratt (2014) concluded, based on a series of ethnographic studies into the application of Zero Vision in the UK construction industry:

> The emergence of Zero Target safety programmes arguably reflects a broader societal desire to quantify and measure human life. . . . The corporate voice of Zero Target speaks of an achievable tangible goal, positioned as a future reality, which can be counted and measured through a plethora of targets. Yet this is challenged and derided by the workforce who position zero as an unachievable target, preferring instead an iconoclastic vision of zero. . . . It is the desire for measurement that brings zero into an ugly reality. Blueprint utopian thinking does not seek to challenge and change current practices. Instead, it aims to operate within the same hostile environment, seeking engagement of the workforce without addressing problems of practice. Furthermore, associations with measurement have arguably encouraged a focus on the numbers and continuous improvement rather than the practices and the people behind them.
>
> (p. 747)

This is probably why studies of Zero Vision are of necessity confounded by a host of other interventions. Just adopting a Zero Vision and doing nothing else (so as not to introduce any confounds), as Sherratt surmised, is likely to generate cynicism and disengagement and not do much good for safety outcomes. It will likely lead to efforts to manage and push a number as low as possible, as some Zero Vision adoptions have devolved into (Long, 2012), which then becomes the empty 'art of managing nothing' (Lofquist, 2010). The deleterious and dehumanising effects we know by now: numbers games, hiding and manipulating injury figures, firing people involved in incidents, not counting fatalities as lost-time injuries. As predicted by Weber over a century ago, this can produce effects that run directly counter to original intentions – in this case, more suffering rather than less.

The studies on Zero Vision suggest that it cannot be used alone and succeed. Zwetsloot et al. demonstrate that it needs to be accompanied by many other things: automation and other technology changes, transformational leadership, adoption of better investigative techniques, change management, introducing a just culture reporting system, and more. However, these same accoutrements needed for the success of Zero Vision eradicate the empirical evidence of its efficacy – and even its necessity. Focus on operations, defer to expertise, become a transformational leader, learn from successes and failures, reduce unnecessary complexity and couplings, rethink accountability relationships, empower the worker, take more risk, and develop risk competency rather than risk averseness. These things all have broad support in the safety literature and don't need Zero. We will look closely at some of them in the next chapter when we discuss what we should measure instead.

Yet, of course, this was true during the Enlightenment. People need a rousing label, a poetic cause, to be stirred into caring and acting. How can a bureaucratic process of management ever rally people behind it? How can increased measurement, paperwork, or compliance inspire people to trust and respect the moral authority it intends to impose? This is where some type of vision (whether 'Zero' or something else) might still be indispensable. But the risk is that it reduces the vision to a banner, a slogan. The real work goes on underneath. And as it does, a Zero vision does not necessarily mean a commitment to zero accidents at all levels of severity. Instead, it might cover severe accidents and imply that near misses and minor accidents are inevitable and essential for learning from complex socio-technical systems' everyday workings and failings. That kind of vision would align with most accident theories, such as standard accident theory, man-made disaster theory, and drift theories. None believe a total zero vision – a world without accidents – is achievable. This goes for high-reliability theory as well.

The rational choice at the frontline

A traditional take on zero harm (that all incidents and injuries are avoidable) entails a surrender to what is known as 'the old view' of human factors or human error: we have a safe system, except for people's choices at the point of risk. This involves the so-called rational choice theory – the premise that people who face a decision choose among fully reasoned alternatives. It keeps the focus on the actions or omissions of frontline operators. 'Unsafe acts,' a term coined by Heinrich in the 1930s to denote exactly this, remains a pivotal concept in the Swiss cheese model widely used today – it separates mere incidents or near misses from actual loss-creating accidents. It reifies the belief that things don't go wrong (however the odds are stacked up) until and unless a frontline worker *does* something wrong.

That makes the target for a Zero vision pretty obvious. Consider the example of a food warehouse, where 150 workers load and unload trucks, lift boxes, drive forklifts, and move pallets. No one reports an injury during the month, and all workers receive prizes, such as $50 gift certificates. If someone reports an injury, no prizes are given that month. Management then added a new element to this 'safety incentive' program: If a worker reported an injury, not only would co-workers forgo monthly prizes, but the injured worker would have to wear a fluorescent orange vest for a week. The vest identified the worker as having a safety problem and reminded co-workers that they lost prizes (Frederick & Lessin, 2000). This is an example of what has been noted in some countries as a neoliberal trend toward worker 'responsibilisation.' A recent Canadian study shows that workers are increasingly blamed (sanctioned, ticketed) for safety violations, with over two-thirds of all citations handed out by workplace safety inspectors directed at them rather than the organisation (Gray, 2009). Workers are 'instructed to become prudent subjects who must "practice legal responsibility"' (p. 327). And if they don't, 'the failure to practise individual responsibility in the face of workplace dangers is often used to explain why workers who perform unsafe jobs become injured' (p. 330).

The premise of full rationality and workers who have their own choices to blame when things go wrong juxtaposes with safety science research.

This generally stresses the influence of context on human action and the role of others in creating the conditions for success and failure in complex systems. Attempts at implementing zero harm keep gravitating toward frontline worker rational choice assumptions. (They 'don't always do what they are supposed to do.') Perhaps these assumptions are hard to root out because they derive from a cultural-historical heritage and a script about why things go wrong and people get hurt that goes much further than that.

The long shadow of choice and suffering

The notion that suffering results from human moral choices has a long historical shadow in the West. Of course, most cultures have evolved allegories about the sources of suffering, which often coincide or are linked with those of their birth. Many start with human beings living in close intimacy with the divine. In a blissful initial state, there is no ontological divide but complete harmony with nature and each other – and no suffering. Storytellers may have invoked these images to reassure people that life was not meant to be so painful, so separated. Then, typically, it follows a separation. The allegory of Adam and Eve, who inhabit the Garden of Eden (placed second among more than 20 creation stories found in the Judeo-Christian bible alone, but likely the oldest one, from around 1000–900 BCE) follows this script. But it does so with a significant distinction from similar contemporary accounts (e.g. the Babylonian epic of Gilgamesh). The Judeo-Christian account places moral responsibility for that separation (and humanity's subsequent introduction to suffering) on the human-on-human responsibility for violating a trusting relationship with the divine (Armstrong, 1996; Visotzky, 1996).

Such, in any case, is the reading by Augustine of Hippo (354–430 CE). His 'theodicy' (or justification of a divine existence despite evil and suffering in the world) answers why we suffer by explaining that evil results from human free will, and sin corrupts essentially good humans. Writing in the early fifth century BCE, Augustine argued that:

> [W]hen an evil choice happens in any being, then what happens is dependent on the will of that being; the failure is voluntary, not necessary, and the punishment that follows is just.

> (Yu, 2006, p. 129)

Suffering, in this reading, is caused by bad human choices; it is the just ret-ribution that follows such choices. Suffering is not inevitable; it hinges on rational human choice. Calvin (1509–1564), instrumental in shaping much of the recent West's interpretation of Judeo-Christian history and ethics, relied heavily on Augustinian theodicy. In *The Bondage and Liberation of the Will* (1543), a publication that mainly addresses the freedom of human will and human choice, Calvin includes many citations from Augustine – signifi-cantly more than from any other patristic authors (e.g. Tertullian, Pelagius), agreeing on the essential links between human choice, sin, and evil.

Zero as ultimate deliverance

If Zero Vision is, in part, a descendant of the Enlightenment (and thus of modernisation and measurement), it may also be a continuation of some of these assumptions and beliefs. Modern, secular institutions (e.g. indus-trial, bureaucratic ones) are webbed with social relations that drive the cre-ation and congealing of 'religious' beliefs, principles, myths, and rituals. Expressions change but do not disappear with modernisation (Douglas, 1992). Such beliefs, principles, moral instructions, myths, and rituals have been noted in health and safety (Besnard & Hollnagel, 2014). As Nietzsche predicted in 1882, religiosity continues to act in corporate and social life (Wood, 2015). In these ways and others:

> [T]he structures of modern industrial society, despite great modifications in different areas and national cultures, produce remarkably similar situations for religious traditions and the institutions that embody these.
>
> (Berger, 1967, p. 113)

Zero harm can take pride of place in this. As Max Weber argued, the alle-viation and redemption of suffering have always been central to religiosity. The whole point of religion – psychologically, socially – was that it supplied rationally constructed systems that help humanity deal with suffering; for Weber, suffering is the driving force behind all religious evolution. At the same time, alleviating suffering is an expressed hope and a call to action, as done in charity and social justice movements. Redemption of suffering is concerned with making suffering somehow meaningful, which religious traditions have done in many different ways – suffering as a test of faith and strength, as a sanction for rule infractions, as a demonstration of humility, as a tutorial for embracing the important things in life, and a lot more. But zero harm is bolder in its aspiration still: it embraces the idea of an ultimate deliverance from suffering. It holds up that 'zero' harm is possible or at least an ideal that organisations can be made to strive for.

So what prescriptions might we follow for developing a Zero Vision for the 21st century? The first makes us question the limits of top-down, rule-driven, centrally governed control over safety outcomes in the pursuit of Zero. The second tells us that we must not focus all our efforts on the frontline, at the point of risk, and tell everyone there (with poster safety campaigns) how to behave. It suggests, thirdly, that we stop looking for a complete deletion of adverse events in our pursuit of Zero. Instead, we might focus on enhancing the positive capacities that make things go well – for which we need to engage differently with what goes on at the frontline. And from that, we may indeed start deriving different kinds of safety 'measures.' Let's turn to those now.

9

WHAT SHOULD YOU MEASURE INSTEAD?

The question of what we should measure instead needs to be preceded by an answer to a different question: *Why* do we want to measure safety? Measuring safety by an absence of adverse events, as we have shown in the previous chapters, doesn't yield much in the way of practical action. It doesn't tell people what to do. In that sense, saying that we measure LTIs (or injury rates) because we need to know what to do is a bit disingenuous. Within the safety professional community, it is broadly accepted that a low number of adverse events does not constitute a meaningful guide to practical action (Besnard & Hollnagel, 2014; Donaldson, 2013; Provan et al., 2017; Rae & Provan, 2019).

That, of course, is not so strange – controlling a system on only one outcome measure (and a very low one at that) can create what is known as the fundamental regulator paradox. The better you control a system, the less that system will tell you. You'll get a zero reading if you are perfect at controlling safety. If you're the person controlling the system, having a zero reading as your only data trace should be quite disconcerting. It doesn't tell you

DOI: 10.4324/9781003177845-9

anything about what you need to do or keep doing, nor does it reveal what might be going on behind that data trace and whether things might blow up any time soon (which, indeed, they have done) (CSB, 2007, 2016). After you have achieved Zero, you literally 'go blind.'

Due diligence

It would be more honest to say that we try to measure safety because of the typical accountability requirements related to the organisation's governance. Safety outcomes are part of reporting expectations (and sometimes requirements). This is particularly the case for listed companies (Dekker, 2022). As we explained earlier, a single homogenised quantity (such as LTI) provides the (supposed) synoptic legibility and easy comparability that markets, and thus boards, are looking for. By this, the hope is that they can demonstrate that they meet their due diligence obligations concerning safety.

Due diligence, however, is not just concerned with receiving information about a number of incidents (e.g. the LTI figure). It isn't satisfied with a superficial understanding of the underlying hazards and risks either. Instead, and importantly, due diligence means you do something with the information you get. You must consider and respond to that information (Tooma & Johnstone, 2012). This is where LTIs or total recordable incidents fall short, as we just discussed. Like any single (very low) outcome measure, they cannot function as a guide for insight – let alone action. The number masks (and may even bend people away from) a consideration and understanding of the actual incidents, hazards, and risks behind it. Yet consideration, understanding, and appropriate action are the essence of due diligence. By their very nature, LTIs are ill suited to inform or support that. Also, LTIs and TRIFR are numbers of injuries – not incidents. An incident may have significant potential but result in no injury. In that respect alone, LTIs and TRIFR fail to support due diligence.

The use of inappropriate metrics does not cause fatalities or disasters. Those causes are found in working conditions, the state of repair of assets, corporate strategic choices, budget allocations, organisational design, structural complexities, goal conflicts, lack of resources, and more. That said, a singular focus on metrics can function as a decoy, taking organisational attention away from the build-up of risks and a possible drift into failure in other areas. Underlying risks can then be left to grow misconstrued or unnoticed, as has been recognised by models of organisational safety since the 1970s (Turner, 1978).

Why things go wrong versus why they go well

But asking why things go wrong and trying to prevent that keeps us locked into a mindset that belongs to a bygone era in safety. Erik Hollnagel (Hollnagel, 2012, 2014b) argued that if we are concerned with safety, we should not (just) try to stop things from going wrong. Instead, we need to understand why most things go well. Because most things do go well. There is much more data to be had – rich data that prevents us from getting trapped in the fundamental regulator paradox. But we seldom know why things go well and what it takes to make it so. (And it's not because everybody is compliant and studiously following all the rules!)

In almost safe systems, we have generally milked the recipes to prevent things from going wrong to the maximum (Amalberti, 2001). Most organisations have many layers of protection in place. They have rules to the point of compliance clutter and internal overregulation. Adding more rules adds more rules. It no longer shows up noticeably in how safe the operations are. Organisations still monitor, record, and investigate the incidents that people get involved in. But the number of things that go wrong is vanishingly small and not a great source of intelligence about the organisation's safety anymore. Safety processes are still driven by something possibly going wrong and organised around stopping it from going wrong. If, following Hollnagel's logic, the organisation can instead get better at finding out what the capacities are that make things go well (which happens a lot!), then it can set about enhancing those capacities and ensuring that as much as possible indeed goes well. It seems obvious.

A spontaneous study in a healthcare organisation that employed 25,000 people (Dekker, 2018b) showed how this can work. The patient safety statistics were dire, if typical: one in thirteen of the patients who walked (or were wheeled) in through the doors to receive care were harmed while receiving that care. That was one in thirteen or 7%. When we asked the health authority what they typically found in the case that went wrong – the one that turned into an 'adverse event,' the one that inflicted harm on the patient – here is what they came up with. After all, they had plenty of data: one out of thirteen in a large healthcare system can add up to a sizable number of patients daily. So, in the patterns that all this data yielded, they found:

- Workarounds
- Shortcuts
- Violations
- Guidelines not followed
- Errors and miscalculations
- Unfindable people or medical instruments
- Unreliable measurements
- User-unfriendly technologies
- Organisational frustrations
- Supervisory shortcomings

It seemed a pretty intuitive and straightforward list. It was also a list that firmly belonged to a particular era in the evolving understanding of safety: that of the person as the weakest link, of the 'human factor' as a set of mental and moral deficiencies that only great systems and stringent supervision can meaningfully guard against. By that logic, the organisation had excellent systems and procedures – it was just the people who were unreliable or non-compliant. As it started dealing with that issue, intervening at the point of risk, modifying people's behaviours at the sharp end, the belief was that numbers would go down (until they reached zero!). Many organisational strategies, to the extent that we can call them that, have indeed been organised around these very premises, as they were in this healthcare system. There were poster campaigns that reminded people of particular risks they needed to be aware of, for instance. And strict surveillance and compliance monitoring concerning certain 'zero-tolerance' or 'red-rule' activities (e.g. hand hygiene, drug administration protocols). A retributive 'just culture' process also got those lower on the medical competence hierarchy more frequently 'just-cultured' (code for suspended, demoted, dismissed, fired) than those with more power in the system. The efforts hadn't paid off. The health authority was still stuck at one in thirteen.

This is when the Hollnagel question started to make sense: 'What about the other twelve? Do you even know why they go well?' The organisation didn't. All its safety and quality improvement resources were directed at investigating, understanding, and trying to prevent what went wrong. There were organisational, reputational, and political pressures to do so, of course, and also presumed legal ones. The resources to investigate the instances of harm were meagre to begin with. The study, such as it was, consisted of some two weeks of field research, observing work as done on or near the frontlines of care delivery. In basically all the patient encounters that led to a good outcome (i.e. not to patient harm), researchers found:

- Workarounds
- Shortcuts
- Violations
- Guidelines not followed
- Errors and miscalculations
- Unfindable people or medical instruments
- Unreliable measurements
- User-unfriendly technologies
- Organisational frustrations
- Supervisory shortcomings

It didn't seem to make a difference! These things always showed up, whether the outcome was good or bad. The safety literature suggests that this is trivial and shouldn't come as a surprise. Vaughan, in a play on Arendt, talked about 'the banality of accidents'; the interior life of organisations is always messy, only partially well coordinated and full of adaptations, nuances, sacrifices, and work that is done in ways that are pretty different from any idealised image of it. When we lift the lid on that grubby organisational life, there is often no discernible difference between the organisation that is about to have an accident or adverse event and the one that won't or the one that just had one (Vaughan, 1999).

This again implies that focusing on people as a problem to control – increasing surveillance, compliance, and sanctioning at the point of risk – does little to reduce the number of negatives. An analysis of 30 adverse events in 380 consecutive cardiac surgery procedures with colleagues in Boston and Chicago (Raman et al., 2016) showed that 30 adverse events occurred despite 100% compliance with the preoperative surgical checklist. These were specific to the nuances of cardiac surgery, the complexities associated with the procedure, patient physiology, and anatomy. Perhaps other adversities were prevented by completely compliant checklist behaviour, even in these 30 cases. But we will never know.

You can also see this in measures of safety culture, which typically include rule monitoring and compliance. They don't predict safety outcomes. One study by Norwegian colleagues conducted in oil production traced a safety culture survey which inquired whether operations involving risk complied with rules and regulations (Antonsen, 2009). The survey also asked whether deliberate breaches of rules and regulations were consistently met with sanctions. The answer to both questions was a resounding 'yes.' Safety on the installation equalled compliance. Ironically, that was a year before that same rig suffered a significant, high-potential incident. Perceptions of compliance may have been significant, but a subsequent investigation showed Vaughan's 'messy interior'; the rig's technical, operational, and organisational planning were in disarray; the governing documentation was out of control; and rules were breached in opening a sub-sea well. Not that these negatives were necessarily predictive of the incident. (Indeed, we need to be wary of hindsight-driven reverse causality.) The messy interior would have been present without an incident happening, too.

Other research in healthcare shows a disconnect between rule compliance, as evidenced in surveys, and how well a hospital keeps its patients safe (Meddings et al., 2017). Hospitals that signed on to a national patient safety project were given technical help – tools, training, new procedures, and other support – to reduce two kinds of infections that patients can get during their hospital stay. One was a central line–associated bloodstream infection (CLABSI) from devices that deliver medicine into the bloodstream. The other was a catheter-associated urinary tract infection (CAUTI) from urine collection devices. Using data from hundreds of hospitals, researchers showed that hospital units' compliance scores did not correlate with how well the units prevented these two infections. As with the oil rig studied by Antonsen, the expectation had been that units with higher scores would do better on infection prevention. They didn't. Some hospitals where scores worsened showed improvements in infection rates. Either way, there appeared to be no association between compliance measurements and infection rates.

Identify and enhance the capacities that make things go well

But if these things don't make a difference between what goes well and what goes wrong, then what does? Hollnagel (2017) argues that what explains the difference isn't the absence of negative behaviours or occurrences (violations, shortcuts, workarounds, and so forth). Instead, the difference lies in the capacities that make things go well (Dekker, 2015). In the 12 cases that went well in the healthcare system, we found more of the following than in the ones that didn't go so well:

- Diversity of opinion and the possibility to voice dissent. Diversity comes in a variety of ways, but professional diversity (as opposed to gender and racial diversity) is the most important one in this context. Yet voicing dissent can be difficult whether the team is professionally diverse or not. It is much easier to shut up than to speak up (Weber et al., 2018). Ray Dalio, CEO of a prominent investment fund, has fired people for not disagreeing with him. He told his employees: 'You are not entitled to hold a dissenting opinion . . . which you don't voice' (Grant, 2016, p. 190).
- Keeping a discussion about risk alive and not taking past success as a guarantee of safety. In complex systems, past results are no assurance of the same outcome today because things may have subtly shifted and changed. Even in repetitive work (landing a big jet, conducting the fourth bypass surgery of the day), repetition doesn't mean replicability or reliability; the need to be poised to adapt is ever present (Woods, 2018). Making this explicit in briefings, toolboxes, and other pre-job conversations that address the subtleties and choreographies of the present task may help things go well.
- Deference to expertise. This is generally deemed critical for maintaining safety. Signals of potential danger, after all, and of a gradual drift into failure can be missed by those unfamiliar with the messy details of practice. Asking the one who does the job at the sharp end rather than the one who sits at the blunt end somewhere is a recommendation from high reliability theory (Weick & Sutcliffe, 2007). Expertise doesn't mean only frontline people. The size and complexity of some operations

can require a collation of engineering, operational, and organisational expertise. Still, high-reliability organisations push decision-
making down and around, creating a recognisable pattern of decisions
'migrating' to expertise.

- Ability to say stop. As Barton and Sutcliffe found in an analysis of incidents,
  '[A] key difference between incidents that ended badly and those that
  did not was the extent to which individuals voiced their concerns about
  the early warning signs' (2009, p. 1339). Edmondson (1999) calls for
  'psychological safety' as a crucial team capacity that allows members
  to speak up safely and voice concerns. In her work on medical teams,
  such capacities were much more predictive of good outcomes than the
  absence of non-compliance or other negative indicators.

- Broken-down barriers between hierarchies and departments. A point
  frequently made in the organisational literature and also in the sociological post-mortems of significant accidents is also one of Deming's
  (1982) reminders, as well as one from the literature on fundamental
  surprises: The totality of intelligence required to foresee bad things is
  often present in an organisation but scattered across various units or
  silos (Woods et al., 2010). Get people to talk to each other: research,
  operations, production, safety, personnel – break down the barriers
  between them.

- Not waiting for audits or inspections to improve. If the team or organisation waited for an audit or an inspection to discover failed parts or
  processes, they were way behind the curve. After all, you cannot inspect
  safety or quality in a process; the people who do the process create
  safety daily. Subtle, uncelebrated expressions of expertise are rife (the
  paper cup on the flap handle of a big jet, the wire tie around the fence so
  the train driver knows where to stop to tip the mine tailings, draft beer
  handles on identical controls in a nuclear power plant control room
  (Norman, 1988)) to know which is which. These are among the kinds
  of improvements and ways workers 'finish the design' of their systems
  so that error traps are eliminated, and things go well rather than badly
  (Petroski, 2018).

- Pride of workmanship. This, according to another of Deming's points, is
  linked to the willingness and ability to improve without being prodded
  by audits or inspections. Teams that take evident pride in the products
  of their work (and the workmanship that makes it so) tended to end up
  with better results. What can an organisation do to support this? They

can start by enabling their workers to do what they want and need to do by removing unnecessary constraints and decluttering the bureaucracy surrounding their daily lives on the job (Germov, 1995).

This list is not a sum of conclusions. It is a set of hypotheses about the input variables that may help produce safe outcomes. These can be starting points for any organisation to identify some of the capacities that make things go well. Let's look at how we might formalise it a little further. Let's go back to the typical safety due diligence requirements that an organisation may need to meet in its governance and oversight and see how some of the research insights from the studies mentioned here could be made to fit – and how this might give rise to new measurements and metrics.

A review of the jurisprudence shows that 'due diligence' for safety has been recognised to have several components (Tooma, 2017):

1. An active and ongoing interest in a baseline of knowledge about safety issues to enable effective decision-making
2. Understanding the nature of the operations that the director is responsible for and the risks that arise from those operations
3. A commitment to addressing those risks through the provision of resources and processes that make managing those risks feasible
4. A proactive approach to seeking out information about incidents, hazards, and risks and considering and responding promptly to that information
5. A commitment to the provision and implementation of processes for meeting relevant duties and obligations
6. Vigilantly verifying the implementation of processes and resources deployed to address risks

Let's look at them individually to get a better sense of what we might want to start measuring to get an index of the capacities that make things go well (Dekker & Tooma, 2022).

The capacity to acquire and maintain safety knowledge

One of the ideas behind this capacity is to gauge to what extent the organisation can anticipate future failure paths. This means monitoring conditions and threats associated with future scenarios within or around the organisation. Anticipation is the expectation of what might happen in the future, which, of course, depends on how we think about the future – and on how we leverage our knowledge of the present and past to inform us about it (Hollnagel, 2017; Weick & Sutcliffe, 2007; Weingart, 1991). In its simplest form, anticipation uses pattern recognition and applies recognition-primed scenario responses to actual or emerging situations (Klein, 1993). Recognition requires suffi-cient similarity between features of known (past) situations and future or present ones so any deductions or inferences have current validity. This may be impossible in complex systems (Cilliers, 2002). Also, predictable kinds of errors intrude on this kind of anticipation, including so-called cognitive garden-pathing or fixation errors (De Keyser & Woods, 1990), in which courses of action are continued as if the scenario is what people anticipated when it has, in fact, subtly changed (Orasanu et al., 1996).

Another possibility for anticipation is the deliberate construction of future scenarios and the preparation of responses to them, sometimes with the use of simulations of various levels of fidelity (Dahlström et al., 2009), which can help people and organisations 'plan for surprise' (Weick, 1995). Anticipation through scenario construction, however, is easier said than done. There is considerable literature and case corpus on fantasy planning (Hutchinson et al., 2018) and fantasy documents (Clarke, 1999). Fantasy documents are artefacts (e.g. response plans, risk assessments) that make optimistic and unrealistic claims (e.g. based on positive audit findings) about how the organisation can control highly uncertain risks to convince stakeholders that the uncontrollable (or, at least, very difficult to control) can be anticipated and bridled (Downer, 2013). Fantasy plans are not usu-ally written to purposefully deceive, although they may have that effect through selective assumptions. Nevertheless, sometimes documents may be written in full knowledge that the stated claims are untrue or have little chance of success. Such claims may be produced by safety departments to persuade a regulator or external auditor or by organisations to persuade public stakeholders. The existence or demonstration of contingency plans, then, is not, in itself, sufficient evidence of the presence of relevant know-ledge. An organisation's ability to track and demonstrate capacity is critical to demonstrating that they are discharging their due diligence duty.

The capacity to acquire and maintain safety knowledge directly relates to Element 1 of the Due Diligence Index-Safety (DDI-S) standard (Titterton et al., 2021). Element 1 is a fundamental building block for effective safety management within organisations. It centres around enhancing the com-prehension of health and safety matters among leaders and executives. This element is pivotal in creating a safety-conscious culture in the organisation. To gauge the effectiveness of Element 1, the key performance indicator (KPI) involves a two-pronged measurement approach. First, it assesses the worker insight experience feedback, weighted by the percentage of worker insights effectively closed out per million hours worked.

Understanding health and safety begins with acknowledging that workers are on the frontlines of operations. They possess unique insights into daily hazards, risks, and operational challenges. Their firsthand experiences make them valuable sources of information when it comes to health and safety. Thus, Element 1 encourages leaders to engage with workers actively, fostering an environment in which insights are valued and considered. The KPI goes beyond just listening to worker insights; it measures the organisation's ability to act on these insights effectively. To do this, organisations calculate the percentage of worker insights that have been successfully resolved or implemented against the total number of insights received. This percentage is multiplied by a million hours worked to establish the insight frequency rate (IFR).

Recognising that it's not just about the quantity of insights, the KPI considers the quality of interactions during safety discussions. After each worker insight activity, the worker and the leader provide ratings on a scale from 1 to 5, with 5 being the highest. Workers are asked questions like 'How was my worker insight experience? On a scale of 1 to 5, was I heard during that experience?' Leaders rate their experiences similarly, evaluating whether they gained valuable insight through the interaction. The KPI combines these ratings to calculate an average percentage score called the worker insight experience feedback rating weighting. This weighted feedback reflects the quality of the relationship between leaders and workers during these safety discussions. It places substantial importance on trust and effective communication between the two groups. This element underscores the idea that building trust and nurturing a positive worker-leader relationship is integral to understanding health and safety.

This element of the standard is pivotal for several reasons:

• *Curiosity and learning*: Element 1 fosters an environment in which leaders are encouraged to be curious about the organisation's health and safety landscape. It instils curiosity and a desire for continuous learning about safety matters. Leaders should actively seek to understand their organisation's health and safety challenges.

- *Direct worker engagement*: The element underscores the value of speaking directly to the task workers. These workers are experts in their roles and the best source of information about hazards and risks. Listening to them is critical to improving safety practices.
- *Implementing practical solutions*: Worker insights often lead to practical and innovative solutions for addressing safety challenges. Organisations can enhance their safety practices and reduce risks by acting on these insights.
- *Quality control feedback*: Element 1 incorporates feedback on the quality of the worker-leader interaction during safety discussions. This feedback loop helps organisations assess whether these activities serve their intended purpose as learning opportunities for health and safety capacity building.
- *Trust and alignment*: The KPI weighting emphasises the importance of trust and alignment between leaders and workers. When both groups share a positive experience during safety discussions, it contributes to a culture in which safety is a shared responsibility.
- *Enhanced safety culture*: Organisations can cultivate a safety-conscious culture by actively engaging workers and valuing their insights. This culture encourages open communication, transparency, and a commitment to continuous improvement in safety practices.

In conclusion, Element 1 is not merely a checkbox in a safety management system. It's a fundamental shift in the way organisations approach health and safety. It encourages leaders to be curious, engage directly with workers, and act on their insights. It places trust and positive interactions at the forefront of safety management. Ultimately, this element is about fostering a workplace where everyone is committed to understanding and improving health and safety. By achieving this, organisations can create safer and healthier environments for their workforce while reaping the benefits of improved safety performance and reduced risks.

The capacity to understand the nature of operations and their risks

Unsurprisingly, understanding the nature of operations and their risks comes from where operations occur and are managed daily (De Carvalho et al., 2009; Havinga et al., 2018). Learning from everyday work and the people who do it is vital to building this capacity. There is always a gap between how work is imagined and done (Hollnagel, 2012). Real work must deal with surprises, unanticipated variations, complications, unpredictable demands, goal conflicts, and resource constraints. People closest to the actual work environment have the most intimate understanding of where the gaps, messy details, and operational nuances are, and they encounter the most opportunities daily to generate and solidify ideas about what can be done to bridge the gaps (Nemeth et al., 2005; Woods et al., 2010).

The aggregate measures – the synoptically legible homogenised quantities of work that we have talked about at length in this book and which boards and senior leaders typically get to see – tend to hide the normal ebbs and flows of strains and shortages that parts of the system are under locally. As a result, when evidence of local adaptations to deal with this first comes to the fore in, for instance, an incident investigation, it tends to get characterised as non-compliance. This won't support learning about daily operations and how their risks get managed quite well despite the pressures and goal conflicts.

Learning from routine work requires organisations to provide opportunities for leaders to engage with work as it is done daily – without a compliance-oriented or judgmental mindset about how things should be done. The ideal is for frontline employees to speak up about their ideas for where, when, and how that work can be improved; what the organisational obstacles are for doing it well; and how compliance pressures and other safety clutter get in the way. This can be done or facilitated through learning teams or reviews – independent of whether work has gone badly, well, or routinely (Havinga et al., 2018; Pupulidy & Vesel, 2017).

Having these activities and processes for learning about routine work in place represents one fundamental way of demonstrating the presence of this capacity (understanding the nature of operations and their risks). The intended targets of this understanding are ultimately the improvements to the design and organisation of work and work environments. These need to happen and be implemented in ways that align with the work as it actually gets done, and they have to deal with the obstacles that generally get in the way of people doing that work (Petroski, 1985; Pew et al., 1981). Such improvements represent a second way of demonstrating the presence of this capacity.

As shown in Chapter 4, worker fatalities are predicted not by (higher) injury rates but rather by a failure on the part of the organisation to understand how they usually create success in their daily work and what sacrifices are necessary to get the job done. Failure to gain that insight has been at the heart of many major disasters, including the 2010 Pike River disaster in New Zealand (where 29 miners lost their lives when a coal mine they were working in exploded). The capacity to understand the nature of the business operations and the risks associated with them should be a fundamental component of any board and senior leadership strategy to minimise the risk of safety disasters in their business.

The capacity to understand the nature of operations and their risks relates to Element 2 of the Due Diligence Index-Safety (DDI-S) standard (Titterton et al., 2021). Element 2 focuses on enhancing leaders' understanding of the organisation's operations and the inherent health and safety risks. Like Element 1, this element serves as a foundational pillar, setting the stage for effectively addressing subsequent due diligence elements.

Before delving deeper into Element 2, it's essential to grasp the associated KPI. This KPI measures the effectiveness of learning teams in increasing leaders' understanding of the organisation's operations and related risks. It assesses the quality of the learning team experience and its impact on enhancing knowledge and awareness of operational aspects that affect health and safety.

This element of the standard is critical for several reasons:

- *Governance structure*: Effective governance structures ensure that leaders at all levels of the organisation are aware of their roles and responsibilities in managing health and safety. They establish clear lines of communication, authority, and accountability. When leaders understand their governance roles in health and safety, they are better equipped to make informed decisions.
- *Alignment with organisational goals*: Element 2 emphasises the importance of aligning safety objectives with broader organisational goals. This alignment ensures that safety isn't viewed in isolation but as an integral part of the organisation's overall strategy. Safety objectives should complement and support achieving the organisation's mission and vision.
- *Reporting systems*: Accountability hinges on transparency and robust reporting systems. Organisations should have mechanisms to report safety incidents, near misses, hazards, and other safety-related issues.

These systems should facilitate timely reporting, tracking, and analysis of safety data.

- *Responsibility and ownership*: Accountability means that individuals and teams take responsibility for ensuring safety. It's about creating a culture in which everyone understands their responsibilities and takes ownership of safety. Leaders are crucial in setting the tone for accountability throughout the organisation.

- *Decision-making integration*: Integrating health and safety considerations into decision-making processes is vital. It ensures that safety is not an afterthought but an inherent part of every decision. This includes considering safety implications when introducing new processes, technologies, or changes in organisational structure.

Element 2 is an integral part of effective safety management in organisations. It underscores the importance of clear governance structures, alignment with organisational goals, robust reporting systems, accountability, and integration of health and safety into decision-making processes. Implementing this element contributes to a strong safety culture, risk mitigation, compliance with regulations, transparency, and continuous improvement in safety practices. By embracing governance and accountability in health and safety, organisations can create safer workplaces, protect their employees, and enhance their overall performance.

The capacity to adequately resource safety

Most organisations don't exist to be safe; they exist to provide a product or a service. Safety may be a precondition for doing so commercially, legally, or ethically, but it is always one of the many goals that need satisfying and achieving. This means that resource battles for safety are likely to be present. As Woods (2003, p. 4) explained,

> [G]oal tradeoffs often proceed gradually as pressure leads to a narrowing focus on some goals while obscuring the tradeoff with other goals. This process usually happens when acute goals like production/efficiency take precedence over chronic goals like safety. If uncertain 'warning' signs always lead to sacrifices on schedule and efficiency, how can any organisation operate within reasonable parameters or meet stakeholder demands?

Goal conflicts work their way down to operational frontlines, where multiple simultaneously active goals are the rule rather than the exception for virtually all domains in which safety plays a role (Boskeljon-Horst et al., 2022). Workers must cope with multiple goals, shifting between them, weighing them, choosing to pursue some rather than others, abandoning one, and embracing another. Many of the goals encountered in practice are implicit and unstated (despite stated priorities for safety). As Hollnagel (2009, p. 94) commented: 'If anything is unreasonable, it is the requirement to be both efficient and thorough at the same time – or rather to be thorough when with hindsight it was wrong to be efficient.' Like efficiency versus thoroughness or safety, other operational and organisational goals often interact and conflict (Dörner, 1989).

Sometimes, these conflicts are quickly resolved in favour of one or another goal, and sometimes they are not. Sometimes, the conflicts are direct and irreducible: for example, when achieving one goal necessarily precludes achieving another – which could, indeed, be safety (Woods et al., 2010). Understanding the nature of these goal conflicts and interactions is crucial if safety is to be resourced adequately in an organisation. A clear line of sight of that trade-off at a board level is crucial to proper decision-making. Resources for a safety organisation should ideally be independent of the organisation's economic performance, and no-jeopardy access to relevant decision-making levels should always be assured (Woods, 1996). However, rather than resourcing the work of safety (e.g. the administrative OHS apparatus, paperwork, processes, and systems), resourcing the safety of work is a much stronger demonstration of commitment to this capacity (Rae & Provan, 2019). It is also a legal requirement in many jurisdictions globally. The next chapter will look at what this might mean for the role of safety people in an organisation.

The capacity to adequately resource safety relates to Element 3 of the Due Diligence Index-Safety (DDI-S) standard (Titterton et al., 2021). Element 3 focuses on the critical aspect of ensuring that organisations have the appropriate resources and processes in place to eliminate or minimise risks to health and safety. This element is essential for the safety and well-being of all individuals in an organisation, and its successful implementation is central to achieving high safety standards. The primary purpose of Element 3 is to bolster the assurance and safety capacity of an organisation's leadership. This is achieved by meticulously assessing the allocated resources and the processes that enable the organisation to address health and safety needs effectively as they arise. In essence, Element 3 seeks to instil confidence in leaders regarding the organisation's readiness to adequately manage health and safety requirements.

The KPI for Element 3 is the '[d]ifference between resources required and resources provided rating.' This KPI measures the discrepancy between the resources deemed necessary for effective health and safety management and the resources that the organisation has provided. It is a tangible metric to assess whether the organisation is adequately equipped to fulfil its safety obligations. Research findings suggest that high-performing organisations possess diverse capacities contributing to superior outcomes. Effective health and safety risk management necessitates organisational capacity, encompassing the right resources, adequate training, relevant skills, and capability to meet the organisation's specific demands. This link between capacity and performance underscores the importance of Element 3, which calls for organisations to invest in capacity assessments.

Furthermore, Element 3 builds on the guidance provided in Element 2 of the DDI-S standard. Organisations must enhance their personnel's capacity and processes to increase the likelihood of positive safety outcomes. Measuring an organisation's capacity-building capabilities is crucial for leadership to gain confidence in the adequacy of organisational resourcing while pinpointing areas that require additional capacity development.

Central to Element 3 is the concept of capacity assessments. These assessments involve a comprehensive evaluation of what an organisation should ideally have in place versus what it currently has in practice. The objective is to determine whether the organisation possesses the necessary resources, which encompass tools and equipment, operational systems and processes, knowledge, and skilled personnel, all required for the safe execution of tasks. A thorough capacity assessment should cover various dimensions, including:

- *Leadership*: Assessing the leadership's role and effectiveness in managing health and safety in the organisation
- *Strategy*: Evaluating the organisation's health and safety strategy and its alignment with the broader organisational goals
- *Structure/governance*: Analysing the organisational structure and governance pertaining to health and safety

- *Resources*: Determining the availability of financial, organisational, and human capital resources for health and safety initiatives
- *Skills*: Assessing the skills and competence of personnel regarding health and safety matters
- *Systems/processes/controls*: Evaluating the effectiveness and implementation of safety systems, processes, and controls
- *Human capital*: Examining the organisation's workforce and its capabilities regarding health and safety
- *Adaptive capacity*: Assessing the organisation's ability to adapt to changing health and safety requirements
- *Accountability*: Evaluating mechanisms of accountability related to health and safety
- *Culture and communication*: Assessing the organisation's culture and communication practices, which is integral to safety performance

When reporting on Element 3, organisations should consider the following:

- Provide an overview of the capacity assessments conducted within the organisation, including details about the tools or methodologies used.
- Explain the survey instruments employed to assess both 'resources as imagined' and 'resources as provided,' along with information on how and to whom these surveys were administered during the reporting period. This should include details on the content of surveys for leaders, subject matter experts, and workers.
- Describe any disparities identified between the resources envisioned and provided as reflected in the average scores obtained from leader, subject matter expert, and worker surveys.
- Compare the organisation's capacity with previous reporting periods to highlight any progress or areas that require attention.
- Outline the specific resources the organisation is allocating to support achieving its desired capacity in health and safety.

By conducting comprehensive capacity assessments and reporting on the alignment between required and provided resources, organisations can pinpoint areas for improvement, foster a robust safety culture, and ensure they have the necessary resources to effectively manage health and safety risks. Furthermore, this process enhances the organisation's overall safety capital, improving morale, well-being, and productivity while minimising hidden costs such as absenteeism.

The capacity to respond to risks and unsafe events

Research suggests that a capacity to deal with risks and unsafe events doesn't typically come from centralised, directed responses but rather from pushing or devolving decision authority down to the points of action and inter- action with the safety-critical process (Loukopoulos et al., 2009; Woods & Patterson, 2001). The adaptive capacity required to deal with risks and unsafe events as they emerge from actual operations can barely be captured in standard protocols or pre-written guidance (Rochlin, 1999; Woods, 1990). Monitoring that capacity offers a critical insight into safety resilience and is a crucial component of the role of boards in overseeing safety. When coupling tightens and interactive complexities escalate (as the saying goes, in a crisis, all correlations go to one), devolving decision authority is known to yield better results in real time – even when horizontal co-ordination is key to preserving overall system safety and integrity (Snook, 2000).

Research shows that adaptive capacity can be grown by emphasising the diversity in the voices of influence and decision-making (Jagtman & Hale, 2007; Janis, 1982; Page, 2007); by letting decisions gravitate toward expertise, not power (Weick & Sutcliffe, 2007); by – as shown in the healthcare study mentioned earlier – instituting and rewarding a willing- ness to say 'stop' even in the face of acute pressures to continue; by allowing operational and design improvements to grow on the frontline without relying on audits or inspections to trigger them; and by encouraging a con- comitant pride of workmanship. These all constitute measurable or at least demonstrable capacities.

Another aspect of demonstrating the capacity to respond is what an organisation does with the people involved in the unsafe event. It has long been known that sanctioning and learning are mutually exclusive and that organisations can do either, but not both simultaneously (Dekker, 2007, 2023). Retributive responses that are organised around rules, violations, and consequences have a way of impeding openness, honesty, and learning. They also don't get to the deeper causes of trouble and tend to fight symptoms instead. The alternative is restorative approaches, in which all stakeholders impacted by an incident or safety event are involved so they can figure out what should be done and by whom to repair the harm done and prevent recurrence.

Because restorative responses ask about the various impacts that an incident or safety event has caused, the needs that arise from those impacts, and whose obligation it is to meet those needs, the kind of accountability they generate is forward looking (Sharpe, 2004): what needs to be done, by whom and when, and how will we know that it is being done? People involved in and affected by the incident collaboratively decide what needs to be done. This can help restore trust between stakeholders, empower victims, and reintegrate practitioners. Restorative justice deals with the consequences and causes of an event. It isn't just between the 'offender' and the 'judge,' and it doesn't pursue narrow facts to secure, for example, a dismissal. This kind of response facilitates a dialogue to identify the many sides of an event and its complex causal web. With a deep understanding of how success is typically assured and how an adverse event could occur, it can create a fair response and identify improvements.

Element 4 of the Due Diligence Index-Safety (DDI-S) standard (Titterton et al., 2021) is centred around the critical aspect of monitoring health and safety in an organisation. It emphasises the need to consider information regarding incidents, hazards, and risks and respond to this information promptly. The primary purpose of Element 4 is to enhance leaders' comprehension of the state of health and safety in the organisation and the critical health and safety risks associated with its operations. It underscores the importance of adopting an inquisitive approach that prioritises learning from work-related experiences. The essence of Element 4 lies in its commitment to fostering a culture of continuous improvement and capacity building in health and safety.

At the heart of Element 4 is a key performance indicator (KPI) that measures an organisation's commitment to monitoring health and safety performance. This KPI is defined by the number of learning reviews and learning team investigations per million hours worked. It is a quantifiable metric to assess an organisation's responsiveness to incidents, hazards, and risks. Monitoring health and safety performance is intricately linked to resourcing investigation processes. This KPI underscores the importance of allocating resources for these investigations. By necessitating learning reviews and learning team investigations, this KPI enables organisations to investigate various incidents ranging from everyday occurrences to serious near-miss and near-hit events. This linkage ensures organisations remain proactive and responsive in their health and safety management approach.

Monitoring health and safety performance in an organisation is a pivotal element for achieving continuous improvement and capacity building. However, the conventional approach to performance assessment in the health and safety context often focuses on outcomes, limiting the depth of understanding. Organisations must embrace a holistic approach that involves investigating incidents, hazards, and risks to overcome this limitation. Organisations are urged to investigate incidents and serious near-miss and near-hit events. These investigations enhance organisational capacity by learning from successes and failures. There exist various methodologies and techniques for effective incident investigation and analysis. The DDI-S standard does not prescribe a specific methodology but requires deploying causal analysis techniques to uncover the root causes of incidents.

Investigating success offers organisations an invaluable opportunity to engage positively with their workforces. Unlike reactive incident investigations, which are traditional, investigating success allows organisations to identify and celebrate what goes right. This process can provide a deeper understanding of effective controls in practice and enables organisations to harness their workforces' positive practices and deviations. Learning teams are a structured approach that empowers organisations to identify and learn from various aspects of routine work. These investigations can encompass factors such as assessing the alignment of 'work as done' with 'work as planned,' understanding the context of deviations, and exploring operational decision-making. Learning teams can be conducted at multiple junctures during operations, including everyday activities; at key milestones in project delivery; at successful project completion; and during the assessment of specific health and safety initiatives.

Organisations are encouraged to adopt a mix of techniques to comprehensively understand the sources of operational success. These techniques answer critical questions, including whether the adopted approach is conducive to health and safety, whether alternatives exist, and how success was derived. Techniques encompass descriptive, normative (norm-finding and norm-comparison), and formative approaches to inquiry. Organisations can gain richer insights into their operations by exploring success from multiple perspectives.

Organisations can leverage the positive investigation methodology (PIM) in the context of serious near-miss and near-hit events. PIM is grounded in the premise that such events represent effective barriers in action. It reframes the traditional view of these events as negative by focusing on what went right to prevent incidents. PIM then proceeds to validate the 'accidental' controls identified in the investigation through a series of 'what if?' questions. These questions test the resilience of effective controls under changing conditions, providing a means of enhancing organisational systems and processes.

By adhering to these principles and guidelines, organisations can foster a culture of continuous improvement in health and safety, engage positively with their workforces, and ensure a proactive approach to monitoring and responding to health and safety information. Element 4 is a cornerstone for building resilience and capacity in health and safety management, ultimately contributing to the organisation's overall success.

The capacity to engage and comply

One of the biggest obstacles in demonstrating engagement is the extent of poor calibration in boards and management (and often even supervisors and workers) about what needs to be complied with (and by whom). Complying with applicable regulations is a minor part of all the compliance demands organisations typically put on themselves and their people (Dekker, 2022). Most compliance demands are internally generated and enforced or expected from business to business (e.g. in a client-contractor relationship) without the relevant regulator knowing or caring (Saines et al., 2014). Many of these rules typically do not correlate with actual legal obligations or safety outcomes but contribute significantly to worker frustration; productivity declines; and, in fact, non-compliance at the front end (Dekker, 2018a).

The amount and putative authority of this safety clutter, however, tends to muddle the organisation's ability to demonstrate compliance with legislation because people inside the organisation (including boards and executives) have a hard time knowing what they are complying with and for whom (Rae et al., 2018). Ideally, a safety measure can be derived from the processes that facilitate engagement and compliance with rules or precautions concerning safety outcomes. For example, unlike the numerous self-imposed rules, there are risk-based processes mandated by regulations that have evolved through bitter experiences on these issues (e.g. confined space, working at heights, or working with hazardous chemicals). This can also be done for

other processes, still demonstrating compliance with the regulatory or legis-
lative requirements (Hale & Borys, 2013a, 2013b; Hale et al., 2013).

Element 5 of the Due Diligence Index-Safety (DDI-S) standard (Titterton
et al., 2021) emphasises establishing organisational processes to ensure
compliance with all health and safety duties and obligations under legislation. This element is designed to assure leaders that their organisation
adheres to health and safety legal obligations across the jurisdictions in which
it operates. The primary purpose of Element 5 is to instil confidence in
organisational leaders that the health and safety management arrangements
comply with the legal obligations in their operational jurisdictions. It serves
as a mechanism to ensure that organisations operate within the bounds of
health and safety laws, thereby reducing the risks and liabilities associated
with non-compliance.

Element 5 is accompanied by a key performance indicator (KPI) that
measures an organisation's commitment to legal health and safety compliance. This KPI is defined by the percentage of legal compliance audit corrective actions closed out. It quantifies the organisation's effectiveness in
addressing corrective actions identified during legal compliance audits.
Organisations' requirement to establish processes for compliance with health
and safety obligations inherently implies the need to conduct health and
safety legal compliance audits. These audits are instrumental in assessing an
organisation's compliance with specific obligations that hold the force of law
under health and safety legislation. The measure associated with Element 5
focuses on reporting the percentage of corrective actions identified during
these audits that have been effectively closed out. This linkage underscores the
organisation's commitment to resources and addresses legal compliance gaps.

Health and safety compliance is not a bureaucratic exercise; it serves a
crucial purpose in safeguarding the well-being of workers and preventing
adverse health and safety outcomes, especially fatal ones. Health and safety
regulations essentially codify control measures derived from industry
practices and lessons learned. These control measures are designed to
mitigate risks effectively and to ensure operational success, particularly in
high-risk situations. Health and safety legal compliance audits differ significantly from other audits conducted to assess conformance with health
and safety management system standards. Legal compliance audits demand a
forensic examination of each legal obligation applicable in the organisation's

jurisdiction(s). The mere performance of a legal compliance audit is not enough to ensure compliance. The results of these audits must be acted on. This entails effectively closing out corrective actions identified during audits to bridge any gaps in compliance. Organisational leaders are pivotal in addressing compliance gaps to achieve legal compliance effectively.

Organisations can demonstrate their commitment to legal health and safety compliance by adhering to these principles and reporting guidelines. Element 5 serves as a critical component in safeguarding the well-being of workers and mitigating the risks associated with non-compliance. Effective legal compliance audits and the closure of corrective actions contribute to a culture of safety and uphold an organisation's due diligence in health and safety management.

The capacity to assure

Practical experience and research on resilience dictate that the control of critical risks, human behaviour, and the control of incident or injury numbers are insufficient to assure safety in a complex system. Safety in complex systems doesn't arise from centralised control and standardisation (which, in the extreme, would outlaw variability) but from acknowledging that variability is inevitable. Guided adaptations to local conditions and challenges will likely generate more significant safety improvements than greater centralised control. (See the next chapter for what that means for the role of safety people.) This depends critically on organisations expanding their adaptive capacity to handle unknown (or even unknowable) disruptions; are they capable of recognising, absorbing, and adapting to harms that fall outside their experience or knowledge base (Hollnagel et al., 2008; Roe, 2013; Sutcliffe & Vogus, 2003)? In other words, controlling adaptive capacity is critical: It is the ultimate demonstration of assurance.

Large organisations contain so many interacting components that the number of things that can go wrong is vast (Dekker et al., 2011). Small events can trigger more extensive failures: outages, leaks, poor performance, and other undesirable outcomes (Dekker, 2011). Whereas controlling critical risks and preventing all possible failure modes are hopeless endeavours in these complex systems, an organisation may still aim to rigorously identify at least some (if not many) of the weaknesses in the system before small failure events trigger them. This can ensure the system is resilient or finds the places it isn't yet. Chaos engineering has been developed as a method of experimentation (in computer infrastructures) that brings systemic weaknesses to light (Rosenthal et al., 2017). It is an empirical process of verification that can lead to more resilient systems and build confidence in the operational behaviour of those systems. Chaos engineering can be as simple as failing one component (even in a simulated setting) and testing how its failure cascades through the organisation. But it can be much more sophisticated: designing and carrying out experiments in a production environment against a small but statistically significant fraction of live operations in a safe-to-fail way. Engagement with these novel pathways to assurance, along with their testing and implementation, represents a strong demonstration of the presence of the capacity. Let's now turn to what that means for what safety people might do in an organisation.

Element 6 of the Due Diligence Index-Safety (DDI-S) standard (Titterton et al., 2021) emphasises the importance of personally and proactively verifying the provision and use of resources and processes outlined in earlier elements. It is vital for organisational leaders to ensure that their health and safety approach aligns with their resource allocation and performance monitoring expectations. The primary purpose of Element 6 is to establish mechanisms for leaders to verify that the organisation's approach to health and safety aligns with their expectations, particularly regarding resource allocation and performance monitoring. This verification process ensures that health and safety efforts align with leadership's vision and organisational goals. One possible way to think about this is by applying something like a safety net promotor score as an index of the health and safety engagement and satisfaction among employees and contractors. This is a way to focus on verification: by measuring the extent of engagement and benefit derived from the organisation's health and safety efforts, its leadership, its contractors, and its employees. This can inform proactive activities, such as conducting learning teams and help explain the conditions contributing to different score responses.

By adhering to these principles and reporting guidelines, organisations can demonstrate their commitment to health and safety engagement and alignment with leadership expectations. Element 6 is pivotal in fostering a safety-conscious culture and ensuring that health and safety efforts meet the organisation's objectives. The safety net promoter score offers a valuable means of assessing engagement and driving continuous improvement in health and safety practices.

10

NOW WHAT SHOULD YOUR SAFETY PEOPLE DO?

In her reflections on what it means to be living under a totalitarian regime, philosopher and political theorist Hannah Arendt delves into the intricate dynamics that this kind of governance produces in the human psyche and human relationships (Arendt, 1967). One of her notable observations was that under such regimes, people tend to exhibit optic compliance, resignation, and cynicism. Optic compliance suggests that people rig up a facade of conformity. (In our world, that would set the stage for 'safety theatre.') People under a totalitarian regime often feign compliance with the system's ideology and rules – not out of genuine belief but as a means of self-preservation. In such regimes, where dissent can be brutally suppressed and conformity is rewarded, people adopt a façade of loyalty and adherence to the regime's principles. This compliance is 'optic' in that it amounts to a superficial act, which is meant to be perceived as conformant by whoever enforces compliance. The inner beliefs, convictions, and values of the people may differ significantly.

The many practices that have sprung up around LTIs (numbers games. creative injury management, hiding incidents and injuries, downgrading or

DOI: 10.4324/9781003177845-10

withholding medical treatment to produce a more acceptable category of injury, firing the worker who had an injury, renaming the event, liberally using 'light' or suitable duties to keep people optically 'working,' rewarding managers or teams with bonuses, denying or downplaying the actual suffering caused by an injury, excluding contractors or fatalities from injury counts, trying to achieve Zero) all both embody and enable optic compliance. If all that is eventually looked at is a single number, then optic compliance gets organised around the favourable presentation of that number. The demand for synoptic legibility that an injury number is supposed to satisfy simultaneously offers a highly convenient route for presenting a version of 'safety theatre' without raising difficult questions. Optic compliance illustrates the extent to which the manufactured insecurity we talked about earlier in the book can become the primary driver of human behaviour. Workers become aware that deviating from the organisation's mandates and noisy commitments (Zero!) can have career or reputational consequences. As a result, they learn to wear a mask of conformity to protect themselves. But by doing so, they inadvertently perpetuate the system they may secretly despise.

It's not as if most people believe or would naturally approve it. That is Arendt's insight, too. Cynicism emerges as a natural byproduct of living within a web of propaganda, surrounded by numbers, games, and deceit. The constant manipulation of truth and reality erodes people's trust in institutions and their fellow workers. As Arendt pointed out, when lies become the norm and truth is systematically distorted, people have no choice but to develop a cynical mindset as a defence mechanism. Cynicism is a coping mechanism that enables people to navigate the complex landscape of misinformation and propaganda and accommodate their roles. It leads to a state of disillusionment in which individuals are sceptical of any information presented to them (which you can regularly see regarding low injury numbers). While this scepticism is initially born out of a desire to protect themselves from ideological manipulation, it can also lead to a broader erosion of social trust and cooperation, making it difficult for workers to contribute more meaningfully to the safety conversation in their organisation.

Indeed, Arendt noted that prolonged exposure to oppression and manipulation eventually leads to a sense of resignation among people. Totalitarian regimes are skilled at systematically dismantling individuals' sense of agency and autonomy. In the face of overwhelming power, people believe that resistance is futile. It causes them to relinquish any belief in the possibility of effecting change. This resignation is not only a psychological response that may make objective sense given the power asymmetries but also a survival strategy. It allows people to conclude that their continued (though unconvinced) participation in perpetuating the state of affairs is their only possibility. Resignation fosters a climate of apathy, in which people become indifferent to the fate of their organisation. By stripping away the belief in their ability to make a difference, any potential opposition is rendered impotent (and may even be ridiculed or morally sanctioned – as happens with those who proclaim they don't believe in 'zero'). Resignation is, in a sense, the psychological culmination of optic compliance and cynicism, as people internalise their perceived powerlessness and cease to challenge the ruling paradigm.

Managing measures

Subduing individual agency, extinguishing genuine belief, and creating a culture of self-preservation and mistrust: if this is what a safety-management regime does in the pursuit of low numbers of negatives (as expressed in LTIs, for instance), it bodes well for neither the people nor the profession. And it won't help improve the safety of their organisation, either. However, how the sense of optic compliance, resignation, and cynicism come through may be rather subtle. Their production tends to be baked into what a safety professional is now typically expected to do. Take the seemingly simple (and attractive) idea of adopting a 'safety culture' (of course, a deeply fraught and scientifically underdeveloped concept) (Henriqson et al., 2014). As many organisations look at it today, a safety culture – like the embrace of Zero – promotes the belief that all incidents and injuries are preventable. Such prevention must be pursued by prioritising safety, identifying hazards, and putting up posters and other reminders to comply with safety requirements. Safety management must be visible across the organisation through ongoing communication, visual materials, and announcements. It can even show up in reactions to incidents or safety-related protests (which may culminate in sudden firings to show how 'important' management believes safety and its safety culture to be) (Edwards & Jabs, 2009; Zaveri, 2020).

A safety professional in today's organisation is someone whose primary purpose is safety management (meaning managing mostly numbers, records, reports, and data) without a core operational purpose for the organisation. As part of this role, safety people are likely involved in the following (Provan et al., 2020):

- *Hazard analysis*: the identification and analysis of factors that might cause operations, or minor aspects of them, to become unsafe.
- *Control implementation*: the application and deployment of interventions (physical, procedural, behavioural) to manage the hazards that were identified. Safety people document these controls through safety management systems, safety plans, safety procedures, and safety rules.
- *Compliance monitoring and surveillance*: ensuring that human actions and work-place realities adhere to the standards expected regarding how hazards should be controlled, including procedural adherence by people on the frontline.
- *Safety campaigns and promotion*: placing posters at or near the point(s) of risk on the frontline and handing out stickers or merchandise with various proclamations of the ruling safety beliefs (such as 'Zero' or 'Incident and Injury Free').

Together, these job aims translate into a range of systems and processes. They can include pre-start safety assessments, job safety analysis (JSA), safe work method statements (SWMS), and permits to work (PTW), for example. They may also involve setting up a system for accommodating the data harvested in surveillance through intelligent vehicle monitoring systems. The objective of all of them is to ensure that frontline employees understand the hazards associated with their work and to create a record of their assent to comply with them, thereby managing the liabilities that would otherwise rest on the organisation.

Compliance, defence, and prevention

Safety people get to monitor compliance with safety risk controls and requirements – ultimately, to prevent safety incidents or injuries that would show up in the LTI count. Safety people also conduct proactive monitoring, including safety audits and behavioural observations. They may be involved in incident investigations to reactively identify and perhaps suggest how to fix controls that were not complied with. Corrective actions are typical outputs of these monitoring activities. They aim to improve safety controls or organisational compliance with them. Safety professionals then help implement and track the completion of corrective actions. This may all show up in the typical processes to generate, communicate, and review reports to make decisions to improve safety. These reports include information about compliance with safety requirements; completion of safety actions (e.g. observations, action closure); and safety incident descriptions, severity, and frequency. This information may sometimes direct safety people to identify the parts of their organisations that require additional management attention and improvement actions.

Safety professionals may have the technical expertise and management experience to facilitate and, where necessary, arbitrate safety decisions between stakeholders. This makes it natural for them to get involved in the discussion, consideration, and decisions required after evidence of an injury and to consider the possibilities for light or suitable duties, thereby managing LTIs. This arbitration can sometimes be required between the organisation's workforce, line management, and third parties (customers, contractors, or regulators). Safety people are often acutely aware of the balance between safety risks, safety compliance requirements, reputational implications, and the 'numbers games' in injury case management. As Hannah Arendt described, the erosion of truth, the growth of cynicism, and a sense of resignation are part and parcel of it.

In addition to these broader (and more profound) implications, there are some immediately apparent consequences of organising safety work along this compliance-focused, centralised, top-down paradigm. Well-adapted activities that smooth over the inevitable contradictions, goal conflicts, resource limitations, and production pressures of real frontline work typically get overlooked (Hollnagel, 2012). They can even retreat from view when a workplace is submitted for formal review, inspection, or audit (McDonald et al., 2002). The reason is that they have no place in how the organisation likes to see itself or believes it needs to be seen by other stakeholders. The organisation and the work of safety that goes on inside it probably lack a vocabulary to accommodate the evidence of such work as done in the first place – other than coarsely (and misleadingly) labelling it as non-compliant behaviour or violations (Dekker, 2018a). To an insider, the expectations and understanding of work – as expressed and upheld in formal notations of that work in procedures, JSAs, or SWMS – never match the reality of what it takes and how actual work gets done. The work becomes merely reactive and performative, once again co-opting the synoptically legible data points the organisation likes to use, but this time deploying them to achieve the safety professional's aims:

> Safety professional activities are defensive in that they seek closure on behalf of the organisation. Safety professionals need to 'tick off' tasks faster than they generate new tasks to avoid being overwhelmed and uncertain about safety risks. An activity that raises more questions than answers generates more new work than it ticks off. Each open item is a personal threat to line management and the organisation since outsiders will see it as a shortfall in safety management. Therefore, there is a strong need to seek closure – ticked boxes, simple answers, and strict processes with well-defined stopping points. Inevitably, this leads to blaming operational units or front-line workers because broader, less-defined answers require broader, less-defined solutions.
>
> (Provan et al., 2020, p. 5)

Safety work organised around achieving favourable outcome numbers that are synoptically legible does not – and cannot – account for the messy details of actual work. It will skip past or over the technical, social, and political complexities of organisational life and the variability of the work done by people in actual workplaces. Pressures and tensions will arise between the work done on the frontlines (and what is necessary to get that work done) and the formal representation of some of the surface features of that work, inevitably leading to disappointments. As said, these can eventually find their way into increasingly ossifying optic compliance, cynicism, and resignation in frontline and other workers and also in the safety people themselves (Dekker & Conklin, 2022).

A centralised, number-driven, controlling mode of safety management can create unintended destructive implications for the safety of people and frontline work (Provan et al., 2017). Safety organised around low numbers of adverse events cannot handle the complexity of managing the risks in modern workplaces. It cannot ever meet the demand for integration of safety management into core operations. Thus, it will fail to meaningfully support safety-related decision-making throughout the organisation (Woods, 2006). With strategies often drifting into failure, there may be an illusion that just a little bit more would be enough – that that would make the crucial difference.

In the history and philosophy of science, of course, such adherence to shared beliefs, values, and methods that form the framework for inquiry refers to a 'paradigm.' Paradigms provide a common understanding of the world, defining the questions to be asked, the methods used, and the criteria for evaluating our findings. Once a paradigm has been established – and safety as the absence of adverse events has been around for a while – we typically tinker with more minor problems created by that paradigm's persistence. Tinkering is designed to fill gaps in existing knowledge within the consensus in the standard model: a little bit more control of what supervisors are doing or how contractors are inducted. Or another safety campaign (more extensive, better this time) that targets the point of risk at the frontline. More robust communication of the safety message. Additional surveillance of work in the field once the technologies make themselves available. These things might be expected from an organisation that still sees safety work as driven by the number of adverse events. Incremental 'more of the same' works to get that number low or lower; however, it cannot move the dial. It is also tone deaf to the cries for change, incapable of hearing or seeing that 'more of the same' has ceased to yield anything new or different.

Safety is the capacity that makes things go well

Suppose we see safety not as the absence of adverse events but as the presence of capacities (as introduced in the previous chapter). This can lead to a qualitative shift (Hollnagel, 2014b). The starting position for safety work then gets organised around the capacities of teams, people, and processes that make things go well, and a very different set of safety roles becomes apparent. These roles are no longer about managing measures or directing and imposing actions on others who do the safety-critical operational work. They are about understanding, knowing, guiding, facilitating, and enabling against the realisation that things go well partly because of the capacity to adapt to the actual circumstances and realities of work as done.

The need to shift away from centralised control becomes evident once we realise that safety is not the simple outcome of a linear process that is somehow synoptically legible. Safe operations (as well as incidents) emerge from different things and processes that act together when exposed to various influences. This is complexity, a characteristic of any system we need to govern today. Complex behaviour arises from the many interactions between the components of a system. This naturally brings penalties, uncertainties, difficulties, and opportunities for surprises and new things. Complex behaviour typically becomes patterned or somewhat repetitive at different levels of analysis or visibility. Probabilities can be predicted in those cases, even if specific outcomes cannot. Complexity asks us to focus not on individual components but on their relationships (Dekker, 2011; Dekker et al., 2011). The properties of the system we might be able to observe emerge as a result of these interactions; they are not contained within individual components. Focusing on the isolated functioning of components and trying to increase their efficiency, reliability, or performance can lead to global inefficiencies and failures. Complex systems generate new structures internally. They are not reliant on an external designer. (They couldn't even be designed.) In reaction to changing environmental conditions, the system will adjust (some of) its internal structure. And this will, in turn, change things in the environment.

Complexity and control

Complexity is a feature of the system in interaction with its environment, not of components inside it. The knowledge of each component is limited and local (which goes for managers and safety people, too!), and no one component possesses the capacity to represent the complexity of the entire system. Decision-makers in complex systems can assess the probabilities but not the certainties of particular outcomes (Orasanu & Martin, 1998). Acquiescing to the notion that systems are complex might give rise to the suspicion that managers stand to lose 'control.' But that misinterprets the science around this:

> During the 1990s and 2000s, authors such as Rasmussen, Woods, Hollnagel, Dekker, Amalberti, and Leveson made increasing calls to pay attention to adaptability as a critical ingredient for safety management. These authors acknowledged the importance of control, but since they were writing when safety management by centralised control was entrenched in organisations, they often positioned their work in contrast to existing practice. This reinforced the widespread perception that control and adaptability could not co-exist. There appeared to be a stark choice between Safety I and Safety II. The mode we present here, 'guided adaptability,' is not a new idea but clarifies that safety comes neither from preventing nor encouraging variation nor recognising that variation is inevitable.
>
> (Provan et al., 2020, p. 6)

Control is possible, but only by relinquishing the traditional idea, the perfect image of it. The goal of safety management and, indeed, the role of safety people is to identify and enhance the capacities that allow a complex system to recognise the need for adaptation and to muster the abilities to do so effectively and safely. Together, these help people approximate the control of a complex system. The acknowledgement of complexity *can* lead to a richer understanding and better control. But there is no procedure for deciding which narrative or control strategy is necessarily the best or the correct one, even though some descriptions and attempts will deliver more interesting results than others – depending, of course, on the goals of those engaged. Complexity invites us to change our posture:

> It tells us that we may not always be able to control or fix a situation. We know that some systems are highly resistant to change. Others may be over-sensitive so that a small interaction may flip the system unpredictably. Rather than seeing such systems in mechanical terms, it would be more effective to feel out and understand how such systems function at an organic level. We need to sense them as living, functioning systems to see how they depend upon complexity levels of meaning so that any action we take flows from an understanding of this underlying meaning.
>
> (Peat, 2002, pp. 152–153)

This is, to an extent, achievable because even in complex systems, it is – at least at some level – not impossible to apply 'laws' or higher-order properties to account for its behaviour. It suggests that complexity and chaos are not things to be afraid of. Instead, we can recognise a great responsibility and opportunity when we realise our connections to each other and the multiple overlapping systems, communities, and societies we are part of, including their innumerable interactions and feedback loops. Because there are exciting things we can do, even things we can foresee. The exact trajectories of parts of the complex system cannot be predicted. But broadly, across a range of behaviours and scales, we may begin to see repetitions and other patterns that at least offer stochastic predictability, allowing us not to predict certainties but to foresee probabilities. Recognise how this allows us to deal with something like Goldbach's conjecture. We can never prove with certainty that *every* even number is the sum of two primes because there are infinite numbers. But by the time we've shown Goldbach to be right for a convincingly large n of even numbers, we could claim that we've established a pretty good pattern, a repetition that allows us to predict the *probability* that the successive two primes we hadn't added up yet – and won't bother to add up, either – will once again yield an even number.

The principles and patterns of the organisation of living systems are unlike those of machines, and we need complexity theory (nonlinear dynamics) to understand their intricacies. Complexity theory tries to explain how simple things can generate complex outcomes that could not be anticipated by just looking at the parts themselves. The systems perspective of living organisations, whose stability is dynamically emergent rather than structurally inherent, means that safety is something a system does, not something a system has (Hollnagel, 2014a; Rasmussen & Svedung, 2000). Failures represent breakdowns in adaptations directed at coping with complexity (Woods, 2003). Resilience represents the system's ability to recognise, adapt to, and absorb a disruption that falls outside the disturbances the system was designed to handle. Making judgments of resilience thus means fixing on higher-order system properties, to leave it 'alive' and not just pick it apart to examine parts.

Safety activities in a complex system

What are some of the activities that would be consistent with the role of safety as guided adaptability in a complex system? Recall from the previous chapter that thinking about safety capacities and due diligence means an active and ongoing interest in a baseline of safety knowledge. This authentic knowledge of work as done enables effective decision-making: it builds a genuine understanding of the nature of the operations, as well as the changing face of risks arising from those operations. It also involves committing to addressing those risks by providing resources and processes for managing them and a proactive approach to sourcing information about whether that is working.

For safety people, a vital capacity is anticipating and predicting (even if only in probability) future failure paths. As said, this requires safety people and managers to have adequate and authentic knowledge of operations. This can lead to considering managerial trade-offs and perhaps even sacrificing decisions designed to counteract eroding safety margins (Cook & Rasmussen, 2005). There are almost always production expectations, which consistently exert pressure on safety margins. As Cook and Rasmussen have shown, maintaining slack is (politically and economically) difficult in organisations. But sacrifice judgments (or trade-offs) may have to be made when operational and financial objectives compromise safety. Sacrifice judgements temporarily relax these acute production or efficiency goals to reduce risks when operations are too close to safety boundaries. There are various ways in which this can be done:

> Safety professionals create, support, and share experiences where safety management is prioritised over production and financial objectives. This situation can be where workgroups have adjusted their work due to emergent safety concerns or additional unbudgeted resources have been provided to preserve safety margins. Celebrating sacrifice judgments as a success encourages managers and employees across the organisation to do the same. Safety professionals celebrate the lost tender because safety was priced in, and the project team went over schedule and over budget to maintain safety margins that were required for unforeseen and, therefore, not planned issues. The organisation sees these as successes for safety, which is very different from other organisations' models of success.
>
> (Provan et al., 2020, p. 11)

Inspired by examples like this, resources can be built up in one part of the operation to beef up adaptive capacity and enhance the ability to respond (Hollnagel, 2018; Woods, 2006). The conditions and emerging threats associated with these scenarios must be monitored across the organisation. This may involve production targets, financial budgets, resource levels, contract requirements, project schedules, and more against the background of assuring business continuity. Safety people can be instrumental in helping with the identification of an organisation's total deployable reserve resources. They can get involved in claiming, negotiating, or deciding to re-distribute human, financial, and technical resources, for example, or assisting local operating units with the authority to requisition additional resources to absorb unexpected demands. Remember, the history of system disasters consistently shows that investing in safety is most important when operational management believes it cannot afford to invest (Vaughan, 1999) or when incident and injury numbers are low (Saloniemi & Oksanen, 1998).

Despite complexity, planning and proactive coordination can still be helpful. An organisation, as explained earlier, doesn't operate entirely in the dark: there are patterns at certain levels of analysis that offer the predictability

of probabilities. These can help direct plans and inform proactive coordination. The knowledge base with which this is done, however, is perpetually
incomplete, which demands that safety people keep searching for new and
emerging information. All plans and risk models are only partially correct;
adaptations will be necessary. Helping others adapt and guiding that adaptation are valuable activities for safety people. To achieve this, safety people
can coordinate and communicate across departments and silos in ways that
operational managers may not be able to. They can identify gaps between
operating units, technical departments, support teams, etc. They also play
an essential role in finding and amplifying the voice from the frontline
(Edmondson, 2019):

> The safety-professional organisation, in part, operates like a shadow, parallel, or
> redundant communication and coordination network throughout the organ-
> isation. Safety information can be exchanged between safety professionals in
> different departments with minimal distortion due to their consistent safety
> vernacular. Safety professionals translate information into ways their local
> operating units and functional departments understand – operations, project
> management, engineering, procurement, finance.
>
> (Provan et al., 2020, p. 10)

As safety people discover more through their activities and movements
throughout the organisation, it becomes clear that enhancing adaptive
capacity itself requires constant adaptation. Rather than planning and
conforming, the guiding notion for the whole organisation has to become
planning and revising. The tracking of a single historical number, as if it
would be a proxy measure of safety, obviously has very little part to play
in this – if any. The roles of safety people and the activities they deploy in
and around those roles are not about managing a single outcome measure,
setting rules, monitoring compliance, and counting what can be counted.
They are about being with sharp-end operators on the frontlines of activity,
studying work as done and the adaptations that entail making or supporting
decisions that reinforce capacities and pushing back on decisions that
undermine adaptive capabilities. They are about coordinating across boundaries, fearlessly probing, and questioning technical specialists, operational
managers, and strategic directions.

REFERENCES

Ali, H., Abdullah, N. A. C., & Subramaniam, C. (2009). Management practice in safety culture and its influence on workplace injury: An industrial study in Malaysia. *Disaster Prevention and Management, 18.*

Alruqi, W. M., & Hallowell, M. R. (2019). Critical success factors for construction safety: Review and meta-analysis of safety leading indicators. *Journal of Construction Engineering and Management, 145*(3).

Amalberti, R. (2001). The paradoxes of almost totally safe transportation systems. *Safety Science, 37*(2–3), 109–126.

Anand, N. (2016, 15 September). Managers must face up to the risk of creating meaningless safety metrics. *TradeWinds,* 21–22.

Angell, I. O., & Straub, B. (1999). Rain-dancing with pseudo-science. *Cognition, Technology and Work, 1,* 179–196.

Antonsen, S. (2009). Safety culture assessment: A mission impossible? *Journal of Contingencies and Crisis Management, 17*(4), 242–254.

Arditi, D., & Chotibhongs, R. (2005). Issues in subcontracting practice. *Journal of Construction Engineering and Management, 131.*

Arendt, H. (1967). *The origins of totalitarianism* (3rd ed.). George Allen & Unwin Ltd.

Armstrong, K. (1996). In the beginning: A new interpretation of Genesis (1st ed.). Alfred A. Knopf. www.loc.gov/catdir/description/random0410/96026170. html

Australian Bureau of Statistics. (2013). *Private sector construction industry, Australia, 2011–12 (No. 8772.0)*. Australian Bureau of Statistics.

Australian Bureau of Statistics. (2016). *Characteristics of employment, Australia, 2015–16 (No. 333.0)*. Australian Bureau of Statistics.

Azari-Rad, H. (2015). Subcontracting and injury rates in construction. *Members-Only Library*, 0(0). http://50.87.169.168/OJS/ojs-2.4.4-1/index.php/LERAMR/article/view/1296

Bahn, S. (2012). Moving from contractor to owner operator: Impact on safety culture – a case study. *Employee Relations, 35*.

Barlas, B. (2012). Shipyard fatalities in Turkey. *Safety Science, 50*. https://doi.org/10.1016/j.ssci.2011.12.037

Barton, M. A., & Sutcliffe, K. M. (2009). Overcoming dysfunctional momentum: Organisational safety as a social achievement. *Human Relations, 62*(9), 1327–1356.

Beck, U. (1992). *Risk society: Towards a new modernity*. Sage Publications Ltd.

Berger, P. L. (1967). *The social reality of religion*. Faber.

Besnard, D., & Hollnagel, E. (2014). I want to believe: Some myths about the management of industrial safety. *Cognition, Technology and Work, 16*(1), 13–23.

Bird, R. E., & Germain, G. L. (1985). *Practical loss control leadership*. International Loss Control Institute.

Blank, V. L. G., Andersson, R., Lindén, A., & Nilsson, B. C. (1995). Hidden accident rates and patterns in the Swedish mining industry due to involvement of contractor workers. *Safety Science, 21*.

Boskeljon-Horst, L., de Boer, R. J., Sillem, S., & Dekker, S. W. A. (2022). Goal conflicts, classical management and constructivism: How operators get things done. *Safety, 8*(37), 1–18.

BP. (2010). *Deepwater Horizon accident investigation report* (British Petroleum, Issue). British Petroleum.

BP. (2017). *HSE charting tool*. BP. http://tools.bp.com/hse-charting-tool.aspx#health-and-safety/

Burnham, J. C. (2009). *Accident prone: A history of technology, psychology and misfits of the machine age*. The University of Chicago Press.

Butz, D., & Leslie, D. (2001). Risky subjects: Changing geographies of employment in the automobile industry. *Area, 33*(2), 212–219. http://www.jstor.org/stable/20004152

Cilliers, P. (2002). Why we cannot know complex things completely. *Emergence, 4*(1/2), 77–84.

Clarke, L. (1999). *Mission improbable: Using fantasy documents to tame disaster*. University of Chicago Press.

Collinson, D. L. (1999). 'Surviving the rigs': Safety and surveillance on North Sea oil installations. *Organization Studies, 20*.

Cook, R. I., & Rasmussen, J. (2005). "Going solid": A model of system dynamics and consequences for patient safety. *Quality & Safety in Health Care, 14*(2), 130–134.

Crush, J., Ulicki, T., Tseane, T., & Veuren, E. J. (2001). Undermining labour: The rise of sub-contracting in South African Gold Mines. *Journal of Southern African Studies, 27*.

CSB. (2007). *Investigation report: Refinery explosion and fire, BP, Texas City, Texas, March 23, 2005* (Report No. 2005–04-I-TX). U.S. Chemical Safety and Hazard Investigation Board.

CSB. (2016). *Investigation report Volume 3: Drilling rig explosion and fire at the Macondo Well, Deepwater Horizon Rig, Mississippi Canyon 252, Gulf of Mexico, April 10, 2010* (11 fatalities, 17 injured, and serious environmental damage REPORT NO. 2010–10-I-OS). U.S. Chemical Safety and Hazard Investigation Board.

Dahlström, N., Dekker, S. W. A., van Winsen, R. D., & Nyce, J. M. (2009). Fidelity and validity of simulator training. *Theoretical Issues in Ergonomics Science, 10*(4), 305–315.

De Carvalho, P. V., Gomes, J. O., Huber, G. J., & Vidal, M. C. (2009). Normal people working in normal organisations with normal equipment: System safety and cognition in a mid-air collision. *Applied Ergonomics, 40*(3), 325–340.

De Keyser, V., & Woods, D. D. (1990). Fixation errors: Failures to revise situation assessment in dynamic and risky systems. In A. G. Colombo & A. Saiz de Bustamante (Eds.), *System reliability assessment* (pp. 231–251). Kluwer Academic.

De silva, N., Samanmali, R., & De Silva, H. (2017). Managing occupational stress of professionals in large construction projects. *Journal of Engineering, Design and Technology, 15*, 00-00. https://doi.org/10.1108/JEDT-09-2016-0066

Dekker, S., Tooma, M., Titterton, A., & Genn, K. (2021). (DDI-S) standard. *Due Diligence Index Council*. https://www.duediligenceindex.online/

Dekker, S. W. A. (2007). *Just culture: Balancing safety and accountability*. Ashgate Publishing Co.

Dekker, S. W. A. (2011). *Drift into failure: From hunting broken components to understanding complex systems*. Ashgate Publishing Co.

Dekker, S. W. A. (2015). *Safety differently: Human factors for a new era*. CRC Press/Taylor and Francis.

Dekker, S. W. A. (2018a). *The safety anarchist: Relying on human expertise and innovation, reducing bureaucracy and compliance*. Routledge.

Dekker, S. W. A. (2018b, August 29). *Why do things go right?* safetydifferently. com.

Dekker, S. W. A. (2022). *Compliance capitalism: How free markets have led to overregulated, unfree workers*. Routledge.

Dekker, S. W. A. (2023). *Stop blaming: Create a restorative just culture*. Independent.

Dekker, S. W. A., Cilliers, P., & Hofmeyr, J. (2011). The complexity of failure: Implications of complexity theory for safety investigations. *Safety Science, 49*(6), 939–945.

Dekker, S. W. A., & Conklin, T. (2022). *Do safety differently*. Pre-Accident Investigation Media.

Dekker, S. W. A., & Pitzer, C. (2016). Examining the asymptote in safety progress: A literature review. *Journal of Occupational Safety and Ergonomics, 22*(1), 57–65.

Dekker, S. W. A., & Tooma, M. (2022). A capacity index to replace flawed incident-based metrics for worker safety. *International Labour Review, 161*(3), 421–443.

Deming, W. E. (1982). *Out of the crisis*. MIT Press.

Derango, J. (2013, April 12). Safety manager receives jail sentence for falsifying records. *OSHA Net: Workplace Safety*. http://washingtonexaminer.com/article/feed/2088502

Donaldson, C. (2013, March 3). Zero harm: Infallible or ineffectual. *OHS Professional*, pp. 22–27.

Dörner, D. (1989). *The logic of failure: Recognising and avoiding error in complex situations*. Perseus Books.

Douglas, M. (1992). *Risk and blame: Essays in cultural theory*. Routledge.

Downer, J. (2013). Disowning Fukushima: Managing the credibility of nuclear reliability assessment in the wake of disaster. *Regulation & Governance, 7*(4), 1–25.

Edmondson, A. C. (1999). Psychological safety and learning behaviour in work teams. *Administrative Science Quarterly, 44*(2), 350–383.

Edmondson, A. C. (2019). *The fearless organisation: Creating psychological safety in the workplace for learning, innovation, and growth*. John Wiley & Sons.

Edwards, M., & Jabs, L. B. (2009). When safety culture backfires: Unintended consequences of half-shared governance in a high-tech workplace. *The Social Science Journal, 46*, 707–723.

FAA. (1996). *National transportation safety board. Public aircraft safety. Safety study NTSB/SS-01/01*. NTSB.

Flight Safety Foundation. (2020). *Measure for measure by Arnold Burnett – Aero safety world*.

Frederick, J., & Lessin, N. (2000). The rise of behavioural-based safety programmes. *Multinational Monitor, 21*, 11–17.

GAO. (2012). *Workplace safety and health: Better OSHA guidance needed on safety incentive programs* (Report to Congressional Requesters, GAO-12-329) (GAO-12-329). G. A. Office.

Geddert, K., Dekker, S., & Rae, A. (2021). How does selective reporting distort understanding of workplace injuries? *Safety, 7*(3), 58. https://doi.org/10.3390/safety7030058

Germov, J. (1995). Medi-fraud, managerialism and the decline of medical autonomy: Deprofessionalisation and proletarianisation reconsidered. *ANZ Journal of Surgery, 31*(3), 51–66.

Giddens, A. (1991). *Modernity and self-identity: Self and society in the late modern age*. Stanford University Press.

Gillen, M., Baltz, D., Gassel, M., Kirsch, L., & Vaccaro, D. (2002). Perceived safety climate, job demands, and coworker support among union and nonunion injured construction workers. *The Journal of Safety Research, 33*.

Graham, B., Reilly, W. K., Beinecke, F., Boesch, D. F., Garcia, T. D., Murray, C. A., & Ulmer, F. (2011). *Deep water: The Gulf oil disaster and the future of offshore drilling* (Report to the President). National Commission on the BP Deepwater Horizon Oil Spill and Offshore Drilling.

Grant, A. (2016). *Originals: How non-conformists change the world*. W. H. Allen.

Gray, G. C. (2009). The responsibilisation strategy of health and safety. *British Journal of Criminology, 49*, 326–342.

Griffin, M. A., & Neal, A. (2000). Perceptions of safety at work: A framework for linking safety climate to safety performance, knowledge, and motivation. *Journal of Occupational Health Psychology, 5*(3), 347–358. https://doi.org/10.1037/1076-8998.5.3.347

Hacking, I. (1990). *The taming of chance*. Cambridge University Press.

Hale, A. R., & Borys, D. (2013a). Working to rule or working safely? Part 2: The management of safety rules and procedures. *Safety Science, 55*, 222–231.

Hale, A. R., & Borys, D. (2013b). Working to rule, or working safely? Part 1: A state of the art review. *Safety Science, 55*, 207–221.

Hale, A. R., Borys, D., & Adams, M. (2013). Safety regulation: The lessons of workplace safety rule management for managing the regulatory burden. *Safety Science, 71*, 112–122.

Hallowell, M., Quashne, M., Salas, R., Jones, M., MacLean, B., & Quinn, E. (2020). *The statistical invalidity of TRIR as a measure of safety performance*. CSRA – The Construction Safety Research Alliance.

Harrison, S., & Dowswell, G. (2002). Autonomy and bureaucratic accountability in primary care: What English general practitioners say. *Sociology of Health & Illness, 24*(2), 208–226.

Haslam, R., Hide, S., Gibb, S., Gyi, D. E., Pavitt, T. C., Atkinson, S., & Duff, R. A. (2005). Contributing factors in construction accidents. *Applied Ergonomics, 36*(4), 401–415.

Havinga, J., Dekker, S. W. A., & Rae, A. J. (2018). Everyday work investigations for safety. *Theoretical Issues in Ergonomics Science, 19*(2), 213–228.

Heinrich, H. W., Petersen, D., & Roos, N. (1980). *Industrial accident prevention* (5th ed.). McGraw-Hill Book Company.

Henriqson, E., Schuler, B., van Winsen, R. D., & Dekker, S. W. A. (2014). The constitution and effects of safety culture as an object in the discourse of accident prevention: A Foucauldian approach. *Safety Science, 70*, 465–476.

Hinze, J., & Gambatese, J. (2003). Factors that influence safety performance of specialty contractors. *Journal of Construction Engineering and Management, 129*.

Hislop, R. D. (1999). *Construction site safety: A guide for managing contractors*. CRC Press.

Hollnagel, E. (2009). *The ETTO principle: Efficiency-thoroughness trade-off. Why things that go right sometimes go wrong*. Ashgate Publishing Co.

Hollnagel, E. (2012, February 22–24). Resilience engineering and the systemic view of safety at work: Why work-as-done is not the same as work-as-imagined (Gestaltung nachhaltiger Arbeitssysteme, 58). In *Kongress der Gesellschaft für Arbeitswissenschaft* (pp. 19–24). Universität Kassel, Fachbereich Maschinenbau.

Hollnagel, E. (2014a). Becoming resilient. In C. P. Nemeth & E. Hollnagel (Eds.), *Resilience engineering in practice: Becoming resilient* (Vol. 2, pp. 179–192). Ashgate Publishing Co.

Hollnagel, E. (2014b). *Safety I and safety II: The past and future of safety management*. Ashgate Publishing Co.

Hollnagel, E. (2017). *Safety-II in practice: Developing the resilience potentials*. Routledge.

Hollnagel, E. (2018). *Safety-II in practice: Developing the resilience potentials*. Routledge.

Hollnagel, E., Nemeth, C. P., & Dekker, S. W. A. (2008). *Resilience Engineering: Remaining sensitive to the possibility of failure*. Ashgate Publishing Co.

Hopkins, A. (2001). *Lessons from Esso's gas plant explosion at Longford*. Australian National University.

Hopkins, A. (2010). *Failure to learn: The BP Texas City refinery disaster*. CCH Australia Limited.

Huang, X., & Hinze, J. (2006). Owner's role in construction safety. *Journal of Construction Engineering and Management*, 132. https://doi.org/10.1061/(ASCE)0733-9364(2006)132:2(164)

Hutchinson, B., Dekker, S. W. A., & Rae, A. J. (2018, May 23–25). Fantasy planning: The gap between systems of safety and safety of systems. In *Australian Safety Critical Systems Association Conference*. Australian Safety Critical Systems Association.

Jagtman, E., & Hale, A. (2007). Safety learning and imagination versus safety bureaucracy in design of the traffic sector. *Safety Science*, 45, 231–251.

Janis, I. L. (1982). *Groupthink* (2nd ed.). Houghton Mifflin.

JICOSH. (1964). Concept of "Zero-accident Total Participation Campaign". Japan International Centre for Occupational Safety and Health, Ministry of Labour.

Jobin, P. (2011). Dying for TEPCO? Fukushima's nuclear contract workers 東京電力　のために死ぬ?　福島の原発請負労働者. *The Asia-Pacific Journal: Japan Focus*, 9.

Johnstone, R., & Quinlan, M. (2006). The OHS regulatory challenges posed by agency workers: Evidence from Australia. *Employee Relations*, 28.

Kenny, B., & Bezuidenhout, A. (1999). Contracting, complexity and control: An overview of the changing nature of subcontracting in the South African mining industry. *Journal of the Southern African Institute of Mining and Metallurgy*, 99.

Kirksey, G. B. (1992). Minimum decencies – A proposed resolution of the pay-when-paid/pay-if-paid dichotomy. *Construction Lawyers, 12.*

Klein, G. A. (1993). A recognition-primed decision (RPD) model of rapid decision making. In G. A. Klein, J. Orasanu, R. Calderwood, & C. E. Zsambok (Eds.), *Decision making in action: Models and methods* (pp. 138–147). Ablex.

Lamare, J. R., Lamm, F., McDonnell, N., & White, H. (2015). Independent, dependent, and employee: Contractors and New Zealand's Pike River Coal Mine disaster. *The Journal of Industrial Relations, 57.*

Laqua, K., Schrader, B., Hoffmann, G. G., Moore, D. S., & Vo-Dinh, T. (1997). Detection of radiation. *Spectrochimica Acta Part B: Atomic Spectroscopy, 52*(5), 537–552.

Lingard, H. C., Hallowell, M., Salas, R., & Pirzadeh, P. (2017). Leading or lagging? Temporal analysis of safety indicators on a large infrastructure construction project. *Safety Science, 91*(2017), 206–220.

Lofquist, E. A. (2010). The art of measuring nothing: The paradox of measuring safety in a changing civil aviation industry using traditional safety metrics. *Safety Science, 48,* 1520–1529.

Long, R. (2012). *For the love of zero: Human fallibility and risk.* Human Dimensions.

Loosemore, M., & Andonakis, N. (2007). Barriers to implementing OHS reforms – the experiences of small subcontractors in the Australian Construction Industry. *The International Journal of Project Management, 25.*

Lorenz, C. (2012). If you're so smart, why are you under surveillance? Universities, neoliberalism, and new public management. *Critical Inquiry, 38*(3), 599–629.

Loukopoulos, L. D., Dismukes, K., & Barshi, I. (2009). *The multitasking myth: Handling complexity in real-world operations.* Ashgate Pub. Ltd.

Macfie, R. (2015). *Tragedy at Pike River mine.* https://nzhistory.govt.nz/media/photo/tragedy-pike-river-mine-rebecca-macfie.

Marcuse, H. (1991). *One-dimensional man.* Routledge.

Maslen, S., & Hopkins, A. (2014). Do incentives work? A qualitative study of managers' motivations in hazardous industries. *Safety Science, 70*(12), 419–428.

Mc Donald, N., Corrigan, S., & Ward, M. (2002). Well-intentioned people in dysfunctional systems. In *5th Workshop on Human Error, Safety and Systems Development.*

Meddings, J., Reichert, H., Greene, M. T., Safdar, N., Krein, S. L., Olmsted, R. N., Watson, S. R., Edson, B., Albert Lesher, M., & Saint, S. (2017). Evaluation of the association between Hospital Survey on Patient Safety Culture (HSOPS) measures and catheter-associated infections: Results of two national collaboratives. *BMJ Quality and Safety, 26*(3), 226 –235.

Mendelhoff, J., & Burns, R. (2013). States with low non-fatal injury rates have high fatality rates and vice-versa. *American Journal of Industrial Medicine, 56*(5), 509–519.

Merton, R. K. (1938). Social structure and anomie. *American Sociological Review, 3*(5), 672–682.

Min, K. B., Park, S. G., Song, J. S., Yi, K. H., Jang, T. W., & Min, J. Y. (2013). Subcontractors and increased risk for work-related diseases and absenteeism. *The American Journal of Industrial Medicine, 56.*

Muzaffar, S., Cummings, K., Hobbs, G., Allison, P., & Kreiss, K. (2013). Factors associated with fatal mining injuries among contractors and operators. *Journal of Occupational and Environmental Medicine, 55.*

Nemeth, C. P., Nunnally, M., O'Connor, M., Klock, P. A., & Cook, R. I. (2005). Getting to the point: Developing IT for the sharp end of healthcare. *The Journal of Biomedical Informatics, 38*(1), 18–25.

Newlan, C. J. (1990). *Late capitalism and industrial psychology: A Marxian critique* (Publication Number 1340534). San Jose State University.

Ng, S. T., Cheng, K. P., & Skitmore, R. M. (2005). A framework for evaluating the safety performance of construction contractors. *Building and Environment, 40.*

Norman, D. A. (1988). *The psychology of everyday things.* Basic Books.

O'Loughlin, M. G. (1990). What is bureaucratic accountability and how can we measure it? *Administration & Society, 22,* 275–302.

O'Neill, S., Martinov-Bennie, N., & Cheung, A. (2013). *Issues in the measurement and reporting of work health can safety performance: A review.* Safety Institute of Australia (NSW Branch, Issue).

Orasanu, J. M., & Martin, L. (1998). *Errors in aviation decision making: A factor in accidents and incidents.* Human Error, Safety and Systems Development Workshop (HESSD).

Orasanu, J. M., Martin, L., & Davison, J. (1996). Cognitive and contextual factors in aviation accidents: Decision errors. In E. Salas & G. A. Klein (Eds.), *Applications of naturalistic decision making.* Lawrence Erlbaum Associates.

Owen, C., Healey, A. N., & Benn, J. (2013). Widening the scope of human factors safety assessment for decommissioning. *Cognition, Technology & Work, 15.* https://doi.org/10.1007/s10111-012-0219-6

Padavic, I. (2005). Labouring under uncertainty: Identity renegotiation among contingent workers. *Symbolic Interactionism, 28.*

Page, S. E. (2007). *The difference: How the power of diversity creates better groups, firms, schools and societies.* Princeton University Press.

Peat, F. D. (2002). *From certainty to uncertainty: The story of science and ideas in the twentieth century.* Joseph Henry Press.

Petroski, H. (1985). *To engineer is human: The role of failure in successful design* (1st ed.). St. Martin's Press.

Petroski, H. (2018). *Success through failure: The paradox of design.* Princeton University Press.

Pew, R. W., Miller, D. C., & Feehrer, C. E. (1981). *Evaluation of proposed control room improvements through analysis of critical operator decisions.* Electric Power Research Institute.

Probst, T. M., & Estrada, A. X. (2010). Accident under-reporting among employees: Testing the moderating influence of psychological safety climate and supervisor enforcement of safety practices. *Accid Anal Prev., 42*(5), 1438–1444. https://doi.org/10.1016/j.aap.2009.06.027

Provan, D. J., Dekker, S. W. A., & Rae, A. J. (2017). Bureaucracy, influence and beliefs: A literature review of the factors shaping the role of a safety professional. *Safety Science, 98,* 98–112.

Provan, D. J., Woods, D. D., Dekker, S. W. A., & Rae, A. J. (2020). Safety II professionals: How resilience engineering can transform safety practice. *Safety Science, 195,* 1067–1080.

Pupulidy, I., & Vesel, C. (2017). The learning review: Adding to the accident investigation toolbox. In *53rd ESReDA seminar.* European Commission Joint Research Centre.

Quinlan, M. (2014). *Ten pathways to death and disaster: Learning from fatal incidents in mines and other high hazard workplaces.* Federation Press.

Quinlan, M., & Bohle, P. (2004). Contingent work and occupational safety. In J. Barling & M. R. Frone (Eds.), *The psychology of workplace safety.* American Psychological Association.

Rae, A. J., & Provan, D. J. (2019). Safety work versus the safety of work. *Safety Science, 111,* 119–127.

Rae, A. J., Weber, D. E., Provan, D. J., & Dekker, S. W. A. (2018). Safety clutter: The accumulation and persistence of 'safety' work that does not contribute to operational safety. *Policy and Practice in Health and Safety*, 16(2), 194–211.

Raman, J., Leveson, N. G., Samost, A. L., Dobrilovic, N., Oldham, M., Dekker, S. W. A., & Finkelstein, S. (2016). When a checklist is not enough: How to improve them and what else is needed. *The Journal of Thoracic and Cardiovascular Surgery*, 152(2), 585–592.

Rasmussen, J., & Svedung, I. (2000). *Proactive risk management in a dynamic society* (pp. 120–132). Swedish Rescue Services Agency. https://dvikan. no/ntnu-studentserver/reports/Risk%20Management%20in%20a%20 Dynamic%20Society.pdf

Rebitzer, J. B. (1995). Job safety and contract workers in the petrochemical industry. *Industrial Relations: A Journal of Economy and Society*, 34.

Rochlin, G. I. (1993). Defining high-reliability organisations in practice: A taxonomic prolegomenon. In K. H. Roberts (Ed.), *New challenges to understanding organisations* (pp. 11–32). Macmillan.

Rochlin, G. I. (1999). Safe operation as a social construct. *Ergonomics*, 42(11), 1549–1560.

Roe, E. (2013). *Making the most of mess: Reliability and policy in today's management challenges*. Duke University Press.

Roquelaure, Y., LeManach, A. P., Ha, C., Poisnel, C., Bodin, J., Descatha, A., & Imbernon, E. (2012). Working in temporary employment and exposure to musculoskeletal constraints. *Occupational Medicine*, 62.

Rosenman, K. D., Kalush, A., Reilly, M. J., Gardiner, J. C., Reeves, M., & Luo, Z. (2006). How much work-related injury and illness is missed by the current national surveillance system? *J Occup Environ Med.*, 48(4), 357–365. https://doi.org/10.1097/01.jom.0000205864.81970.63

Rosenthal, C., Hochstein, L., Blohowiak, A., Jones, N., & Basiri, A. (2017). *Chaos engineering: Building confidence in system behaviour through experiments*. O'Reilly.

Roughton, J. E. (1995). Contractor safety. *Professional Safety*, 40.

Rousseau, D. M., & Libuser, C. (1997). Contingent workers in high-risk environments. *California Management Review*, 39.

Sætren, G. B., & Laumann, K. (2015). Effects of trust in high-risk organisations during technological changes. *Cognition, Technology & Work*, 17.

Safe Work Australia. (2014a). *Work-related traumatic injury fatalities, Australia 2012–2013*. Released July 2014.

Safe Work Australia. (2014b). *Work-related traumatic injury fatalities, Australia 2009–2010*. Released March 2012.

Safe Work Australia. (2014c). *Work-related traumatic injury fatalities, Australia 2014–2015*. Released March 2015.

Saines, M., Strickland, M., Pieroni, M., Kolding, K., Meacock, J., Nur, N., & Gough, S. (2014). Get out of your own way: Unleashing productivity. In G. Vorster, C. Richardson, & D. Redhill (Eds.), *Building the lucky country: Business imperatives for a prosperous Australia*. Deloitte Touche Tohmatsu.

Salas, R., & Hallowell, M. (2016). Predictive validity of safety leading indicators: Empirical assessment in the oil and gas sector. *Journal of Construction Engineering and Management, 142*, 04016052.

Salminen, S., Saari, J., Saarela, K. L., & Räsänen, T. (1992). Fatal and non-fatal occupational accidents: Identical versus differential causation. *Safety Science, 15*(2), 109–118. https://doi.org/10.1016/0925-7535(92)90011-N

Saloniemi, A., & Oksanen, H. (1998). Accidents and fatal accidents: Some paradoxes. *Safety Science, 29*, 59–66.

Saul, J. R. (1993). *Voltaire's bastards: The dictatorship of reason in the West*. Vintage Books.

Sawacha, E., Naoum, S., & Fong, D. (1999). Factors affecting safety performance on construction sites. *The International Journal of Project Management, 17*.

Scott, J. C. (2012). *Two cheers for anarchism*. Princeton University Press.

Sharpe, V. A. (2004). *Accountability: Patient safety and policy reform*. Georgetown University Press.

Sherratt, F. (2014). Exploring 'zero target' safety programmes in the UK construction industry. *Construction Management and Economics, 32*(7–8), 737–748.

Sherratt, F., & Dainty, A. R. J. (2017). UK construction safety: A zero paradox. *Policy and Practice in Health and Safety, 15*(2), 1–9.

Simon, J. M., & Piquard, P. (1991). Contractor safety performance significantly improves. In *Paper presented at the SPE health, safety and environment in oil and gas exploration and production conference*.

Smecko, T., & Hayes, B. (1999). Measuring compliance with safety behaviors at work. In 14th *Annual conference of the society for industrial and organizational psychology*.

Snook, S. A. (2000). *Friendly fire: The accidental shootdown of US Black Hawks over Northern Iraq*. Princeton University Press.

Størkersen, K. V., Antonsen, S., & Kongsvik, T. (2017). One size fits all? Safety management regulation of ship accidents and personal injuries. *Journal of Risk Research*, 20(9), 1154–1172. https://doi.org/10.1080/13669877.201 6.1147487

Suruda, A., Whitaker, B., Bloswick, D., Philips, P., & Sesek, R. (2002). Impact of the OSHA trench and excavation standard on fatal injury in the construction industry. *Journal of Occupational and Environmental Medicine/ American College of Occupational and Environmental Medicine*, 44, 902–905. https://doi.org/10.1097/00043764-200210000-00007

Sutcliffe, K., & Vogus, T. (2003). Organising for resilience. In K. S. Cameron, I. E. Dutton, & R. E. Quinn (Eds.), *Positive organisational scholarship* (pp. 94–110). Berrett-Koehler.

Taylor, A. (2023). *The age of insecurity: Coming together when things fall apart* (The CBC Massey Lectures). House of Anansi Press.

Taylor, F. W., United States Congress House, & Special Committee to Investigate the Taylor Other Systems of Shop Management. (1926). *Testimony of Frederick W. Taylor at hearings before special committee of the house of representatives January 1912: A classic of management literature reprinted in full from a rare public document.* Taylor Society.

Thebaud-Mony, A., Levenstein, C., Forrant, R., & Wooding, J. (2011). *Nuclear servitude*. Routledge.

Tooma, M. (2017). *Safety, security, health and environment law* (2nd ed.). The Federation Press.

Tooma, M., & Johnstone, R. E. (2012). *Work health and safety regulation in Australia: The model act.* The Federation Press.

Townsend, A. S. (2013). *Safety can't be measured.* Gower Publishing.

Tozer, J., & Hargreaves, H. (2016). *Lost miners: The tragic toll of FIFO work.* SBS.

Turner, B. A. (1978). *Man-made disasters.* Wykeham Publications.

Uher, T. E. (1991). Risks in subcontracting: Subcontract conditions. *Construction Management and Economics*, 9.

Valluru, C. T., Rae, A., & Dekker, S. (2020). Behind subcontractor risk: A multiple case study analysis of mining and natural resources fatalities. *Safety*, 6(3), 40. https://doi.org/10.3390/safety6030040

Vaughan, D. (1999). The dark side of organisations: Mistake, misconduct, and disaster. *Annual Review of Sociology*, 25(1), 271–305.

Visotzky, B. L. (1996). *The genesis of ethics* (1st ed.). Crown Publishers.

Watts, A. W. (1951). *The wisdom of insecurity: A message for an age of anxiety.* Vintage Books

Weber, D. E., MacGregor, S. C., Provan, D. J., & Rae, A. R. (2018). "We can stop work, but then nothing gets done." Factors that support and hinder a workforce to discontinue work for safety. *Safety Science, 108,* 149–160.

Weick, K. E. (1995). *Sensemaking in organisations.* Sage Publications.

Weick, K. E., & Sutcliffe, K. M. (2007). *Managing the unexpected: Resilient performance in an age of uncertainty* (2nd ed.). Jossey-Bass.

Weingart, P. (1991). Large technical systems, real-life experiments, and the legitimation trap of technology assessment: The contribution of science and technology to constituting risk perception. In T. R. LaPorte (Ed.), *Social responses to large technical systems: Control or anticipation* (pp. 8–9). Kluwer.

Wood, M. (2015). Shadows in caves? A re-assessment of public religion and secularisation in England today. *European Journal of Sociology, 56*(2), 241–270.

Woods, D. D. (1990). Risk and human performance: Measuring the potential for disaster. *Reliability Engineering and System Safety, 29*(3), 387–405.

Woods, D. D. (1996). Automation: Apparent simplicity, real complexity. In M. Mouloua & R. Parasuraman (Eds.), *Human performance in automated systems: Current research and trends* (pp. 1–7). Erlbaum.

Woods, D. D. (2003). *Creating foresight: How resilience engineering can transform NASA's approach to risky decision making.* US Senate Testimony for the Committee on Commerce, Science and Transportation, John McCain, Chair.

Woods, D. D. (2006). How to design a safety organisation: Test case for resilience engineering. In E. Hollnagel, D. D. Woods, & N. G. Leveson (Eds.), *Resilience engineering: Concepts and precepts* (pp. 296–306). Ashgate Publishing Co.

Woods, D. D. (2018). The theory of graceful extensibility: Basic rules that govern adaptive systems. *Environment Systems and Decisions, 38,* 433–457.

Woods, D. D., Dekker, S. W. A., Cook, R. I., Johannesen, L. J., & Sarter, N. B. (2010). *Behind human error.* Ashgate Publishing Co.

Woods, D. D., & Patterson, E. S. (2001). How unexpected events produce an escalation of cognitive and coordinate demands. In P. A. Hancock & P. Desmond (Eds.), *Stress, workload and fatigue* (pp. 290–304). Lawrence Erlbaum Associates.

Yu, X. (2006). Understanding suffering from Buddhist and Christian perspectives. *Christian Study Centre on Chinese Religion and Culture*, 7(1–2), 127–152.

Zadow, A. J., Dollard, M. F., McLinton, S. S., Lawrence, P., & Tuckey, M. R. (2017). Psychosocial safety climate, emotional exhaustion, and work injuries in healthcare workplaces. *Stress and Health*, 33(5), 558–569. https://doi.org/10.1002/smi.2740

Zaveri, M. (2020, May 4). An Amazon Vice President quit over firings of employees who protested. *New York Times*, p. 3.

Zohar, D. (1980). Safety climate in industrial organisations: Theoretical and applied implications. *The Journal of Applied Psychology, 65.*

Zohar, D., & Luria, G. (2003). The use of supervisory practices as leverage to improve safety behaviour: A cross-level intervention model. *The Journal of Safety Research, 34.*

INDEX

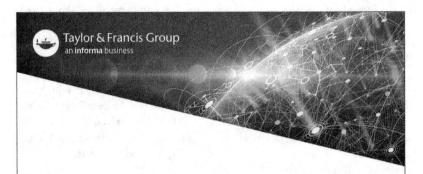

Printed in the United States
by Baker & Taylor Publisher Services